AF576044

Shale Oil and Shale Gas Resources

Shale Oil and Shale Gas Resources

Special Issue Editors

José A. Torres
Hector Klie

MDPI • Basel • Beijing • Wuhan • Barcelona • Belgrade • Manchester • Tokyo • Cluj • Tianjin

Special Issue Editors
José A. Torres
University of Pau and Pays de l'Adour
France

Hector Klie
Rice University
USA

Editorial Office
MDPI
St. Alban-Anlage 66
4052 Basel, Switzerland

This is a reprint of articles from the Special Issue published online in the open access journal *Energies* (ISSN 1996-1073) (available at: https://www.mdpi.com/journal/energies/special_issues/shale_oil).

For citation purposes, cite each article independently as indicated on the article page online and as indicated below:

LastName, A.A.; LastName, B.B.; LastName, C.C. Article Title. *Journal Name* **Year**, *Article Number*, Page Range.

ISBN 978-3-03928-875-5 (Pbk)
ISBN 978-3-03928-876-2 (PDF)

© 2020 by the authors. Articles in this book are Open Access and distributed under the Creative Commons Attribution (CC BY) license, which allows users to download, copy and build upon published articles, as long as the author and publisher are properly credited, which ensures maximum dissemination and a wider impact of our publications.
The book as a whole is distributed by MDPI under the terms and conditions of the Creative Commons license CC BY-NC-ND.

Contents

About the Special Issue Editors

José A. Torres is a Senior Research Engineer with international experience in R&D projects focused on subsurface technologies. His principal areas of interest include different aspects related to reservoir engineering, such as reservoir simulation, production mechanisms, and enhanced oil recovery, fluid flow in porous media, thermodynamics of reservoir fluids, and coupled multiphase flow and geomechanics. Since November 2018, he has been working on R&D activities for advancing geological carbon dioxide storage within the Computational Hydrocarbon Laboratory for Optimized Energy Efficiency at the University of Pau, France. Prior to this role, he was a Senior Reservoir Engineer–Unconventional Reservoirs, at the University of North Dakota's Energy & Environmental Research Center (2016–2018), Senior Reservoir Engineer at ConocoPhillips' Reservoir Engineering Technology organization (2012–2015), Research Engineer at ADERA (2007–2011), and Research Engineer at PDVSA-Intevep (2000–2003). He has published numerous papers at national and international conferences in the areas of reservoir simulation, thermal, and CO2 EOR, CO2 storage, and advanced reservoir monitoring techniques. Dr. Torres completed his Ph.D. in Chemical and Process Engineering at Rovira i Virgili University, 2008, Spain. He also completed a specialist degree in Reservoir Engineering (2002), and graduated as B.S. in Chemical Engineering (1999), both of them at Simón Bolívar University.

Hector Klie is an experienced computational and data scientist focused on developing physics-informed AI solutions for multiple engineering and geoscientific applications in oil and gas. Dr. Klie was co-founder and has been Chief Executive Officer of DeepCast.ai since June 2017. He is currently appointed as an Adjunct Professor in the Department of Computational and Applied Mathematics at Rice University. Prior to his role at DeepCast.ai, he was the Director of Enterprise Data Solutions and Data Science Technical Lead at Sanchez Oil & Gas Corporation (2016–2017), Staff Reservoir Engineer and Lead Data Scientist at ConocoPhillips (2008–2016), Associate Director and Senior Research Associate at the Center for Subsurface Modelling in The University of Texas at Austin (2003–2008), and Research Scientist at PDVSA-Intevep (1989–2003). He has published over 80 papers in the areas of sparse linear solvers, production forecasting, field optimization, uncertainty quantification, high-performance computing, reduced-order modelling, and machine learning. Dr. Klie has patented 5 inventions in the areas of data analytics for automated drilling and parallel physics-based solvers. He has chaired and co-organized several technical events at SPE, SIAM, SEG, and IEEE. He is currently an Associate Editor of the *Computational Geosciences Journal*. Dr. Klie completed his Ph.D. in Computational Science and Engineering at the Department of Computational and Applied Mathematics at Rice University, 1997, and a M.S. in Computer Science at Simon Bolivar University, Venezuela, 1991.

Preface to "Shale Oil and Shale Gas Resources"

The past decade has witnessed incredible production growth from shale oil and shale gas resources, which has reshaped the petroleum industry landscape in many ways. In a remarkable turn of events, the success of the North American shale industry has become one of the most important milestones for the industry in recent decades. The great achievements of the U.S. shale players serve as an outstanding example that has generated considerable interest around the world.

However, despite the operational success to date, many issues are not yet completely understood. Among other factors, the complex nature of rock–fluid interactions on fractured media, together with the relatively short production span, demands novel technological solutions for optimizing drilling, completion, and improved recovery efficiencies.

This Special Issue, entirely devoted to shale oil and shale gas resources, expands on essential technical challenges for improving the understanding and management of these unconventional reservoirs. The idea is to provide an instrument for sharing information and lessons learned, collected from different plays. We hope that this Special Issue will foster fruitful debates among contrasting perspectives toward optimal shale play management.

José A. Torres, Hector Klie
Special Issue Editors

Article

Organic Matter Pore Characterization of the Wufeng-Longmaxi Shales from the Fuling Gas Field, Sichuan Basin: Evidence from Organic Matter Isolation and Low-Pressure CO_2 and N_2 Adsorption

Zhuo Li [1,2,*], Zhenxue Jiang [1,2], Hailong Yu [1,2] and Zhikai Liang [1,2]

1 State Key Laboratory of Petroleum Resources and Prospecting, China University of Petroleum, Beijing 102249, China; Jiangzhenxue0405@163.com (Z.J.); yhl2017210211@163.com (H.Y.); Liangzhikai2020@163.com (Z.L.)

2 Unconventional Natural Gas Research Institute, China University of Petroleum, Beijing 102249, China

* Correspondence: zhuo.li@cup.edu.cn; Tel.: +86-010-8973-9051

Received: 15 January 2019; Accepted: 26 March 2019; Published: 28 March 2019

Abstract: Organic matter (OM) pores are significant for shale gas accumulation and flow mechanisms. The pores of Wufeng-Longmaxi (W-L) shale in the Sichuan Basin, China have been extensively characterized, however, the proportion of OM pores in this shale have not been adequately discussed. In this study, the contribution of OM pores to the total pore volume of W-L shale was quantitatively studied through the analysis of OM isolation, field emission scanning electron microscopy (FE-SEM) and low-pressure CO_2 and N_2 adsorption (LPGA). FE-SEM images showed abundant OM pores, interparticle pores and intraparticle pores with various shapes and widths in the W-L shales. The pore size distribution (PSD) of the isolated OM from five shale samples showed a consistent, unimodal pattern. The pore volume of isolated OM was greater than that of the bulk shale samples, suggesting that OM is more porous than the inorganic compositions in shales. The average contribution of OM to the volumes of micropores, mesopores and macropores was 58.42%, 10.34% and 10.72%, respectively. Therefore, the pore volume of the W-L shale was dominantly related to inorganic minerals. This was probably due to the small weight ratio of OM in the shale samples (1.5 wt%–4.2 wt%). The findings of this study reveal the different effects of OM and minerals on pore development, and provide new insights into the quantitative contribution of OM pores to the total pore volume of the W-L shale.

Keywords: isolated organic matter; organic matter pores; pore size distribution; Wufeng-Longmaxi shale; fuling gas field

---

1. Introduction

Organic shales commonly contain complex pore systems with various pore types, pore geometry and multiscale pore widths [1–3]. The pores in shale are commonly classified as interparticle (interP), intraparticle (intraP), and organic matter (OM) pores or microfractures, using direct imaging techniques [2]. In addition, the pore size of shale can be sorted. The categories include micropore (pore width smaller than 2 nm), mesopore (pore width between 2 and 50 nm) and macropore (pore width larger than 50 nm), following the international union of pure and applied chemistry (IUPAC) classification [4]. Due to the nanoscale organic and inorganic pores within the shale formations, the solid–fluid interaction becomes non-trivial, causing the gas flow to derivate from classical Darcy's law [5,6]. Recently researchers have reported some pore-scale studies of fluid flows in shale, however, most of them focused on a single nano-tube with the mean pore-throat radius. All the derived formulas are based on gas molecular dynamics, which might not be applicable to shale formations, especially the organic pores. Therefore, molecular dynamic simulations are also used to investigate the gas state and flow mechanisms within organic pores [7,8]. Assuming shale formations to be fractal porous media, non-Darcy flows were further investigated. However, the real pore structures within the shale formations are much more complex than the studies stated. Naraghi and Javadpour (2015) first proposed their stochastic model by representing organic patches randomly scattered within inorganic pores [9], and this idea was further adopted by Naraghi et al. (2018) [10,11]. Therefore, characterization of the pores in gas shale is critical to understanding gas storage and gas flow mechanisms [12].

Pores in shale can be characterized by direct observation [2,13–17], radiation detection [3,18–20], and fluid intrusion techniques [1,21–24]. Recently, the multi-scale pore structure (e.g., pore types, shapes, size and connectivity) of shales has been studied extensively using various methods [20]. For example, field emission-scanning electron microscopy (FE-SEM) [2,13–17], small angle neutron scattering (SANS) [3,19,20], nuclear magnetic resonance (NMR) [25–29], nano-scale X-ray computed tomography (Nano-CT) [18,30–32], low-pressure gas physisorption (LPGA) and mercury intrusion capillary pressure (MICP) [1,21–24,33–38] were applied to characterize the pores in shales. Of these techniques, FE-SEM and Nano-CT were effectively applied to the quantitative characterization of OM pores, with the help of image processing, data extraction and electron tomography of subnanometric resolution to characterize the porous network in organic matter [13–18,30–32]. However, these measurements have the drawbacks, the tiny sample preparation for Nano-CT and the limited observation areas for FE-SEM. The results of microscopic observation methods, considering the heterogeneity of organic shales and the small observation area, may not be representative [31]. In addition, OM pores were poorly characterized with the fluid intrusion (LPGA and MICP) techniques because these methods cannot distinguish OM from inorganic mineral compositions [35]. Rexer et al. (2014) proposed an effective OM isolation process that does not alter the pore structure [39]. Therefore, OM isolation from bulk shale provides an effective way to characterize OM pores [39–41].

OM pores are proposed to be the significant pore type in shale [2,13,15–17] and the major contributor to the gas storage capacity for both adsorbed gas and free gas [42–46]. Porosity within OM particles could be larger than 40% based on SEM images [13,15,16,47,48]. OM porosity is promoted by high total organic carbon (TOC) content and appropriate thermal maturity, whereas excessive maturity may reduce the number of OM pores [49–51]. The contribution of OM pores is suggested to be significant in many gas shales [2,13]. Loucks et al. (2012) estimated the proportion of OM pore to porosity of shales using SEM images [2]. Tian et al. (2013) calculated the contribution of OM to pores according to the regression line of TOC content versus porosity [3]. Considering that the pore structure of gas shales may also be controlled by inorganic minerals [13,14,17,31,46–48,50], the accuracy of the results of Loucks et al. (2012) and Tian et al. (2013) may be low [41].

The pore characteristics of Wufeng-Longmaxi (W-L) shale in China were studied by Chinese scholars using various techniques. Quantitative pore structure parameters (i.e., pore size distribution, pore volume and surface area, and pore connectivity) were studied by fluid intrusion methods [20,33–37,40,42,52–54]. OM pore characterization of W-L shale in China has gained increasing research attention and OM types and OM pore networks have been qualitatively investigated by the FE-SEM imaging technique [14,31,32,35,36]. Ji et al. (2017) reported that OM pores in W-L shale were characterized using isolated OM samples [40]. However, they did not quantitatively study the proportions of OM pores. The contribution of OM pores to the total pore volume of the W-L shale requires further study. Therefore, the objectives of this study are to: (1) prepare isolated OM samples from bulk W-L shale, (2) compare the pore characteristics of isolated OM and the corresponding bulk shale using FE-SEM and LPGA (CO_2 and N_2) methods, and (3) quantitatively investigate the contributions of OM and mineral compositions to the pore volumes of the W-L shale. The findings of this paper will provide new insights into the quantitative contribution results of OM and mineral compositions to the total pore volume of W-L shale.

2. Materials and Methods

2.1. Materials

A total of five W-L core samples were collected from well J-4 in the Fuling gas field, Sichuan Basin, China. The geological settings of the Wufeng-Longmaxi Formation in the study area have been reported in previous works [36,42]. Briefly, this field is situated in the southeast margin of the Sichuan Basin, which has experienced multiple stages of tectonic movements, including the Sinian-Silurian Caledonian, Devonian-Permian Hercynian, Triassic Indosinian, Jurassic-Cretaceous Yanshanian, and Tertiary-Quaternary Himalayan movements [36]. The W-L organic shale in the study area is mainly deposited in deep water shelves with low-energy and anoxic environments [41]. The graptolite-rich W-L shale, with a present-day thickness of about 60–110 m, consists of black carbonaceous shale, carbonaceous mudstone, argillaceous siltstone, and siliceous shale [36]. The shale samples in this study were selected from the bottom of the W-L Formation.

The core samples were prepared as rock chips for FE-SEM and crushed to 60 mesh for OM isolation, and 100 mesh for TOC, XRD, and LPGA (CO_2 and N_2) measurements. The particle size used for gas adsorption porosimetry strongly influences porosimetric results and the 60–140 mesh particle size of organic shale is recommended for LPGA experiments [55,56].

2.2. Methods

2.2.1. OM Isolation

OM (mixture of kerogen and bitumen) samples were prepared using the chemical treatments proposed by Rexer et al. (2014) [39]. Specifically, each shale sample, with a particle size of 100 mesh, was first treated with HCl for about 24 h to remove carbonates. After washing several times with distilled water, the residue was treated with HF for 12 h to remove silicate minerals. The residual solids were washed and treated with $CrCl_2$ for 12 h to remove pyrite, and then washed again with distilled water. The OM samples were separated via filtering. The OM samples were prepared after the residues were washed with distilled water and dried at −5 °C. The bitumen in the W-L shale was preserved, considering that abundant bitumen pores exist in more mature shales [31,42].

2.2.2. TOC and XRD

TOC and XRD measurements were conducted following the Chinese national standard (GB/T 31483-2015). Specifically, shale samples of 100 mesh were treated with HCl (10%) to remove carbonates, washed with distilled water and dried for 24 h at 70 °C before the TOC experiments commenced. The TOC content was measured on a Leco CS 230 carbon/sulfur analyzer (LECO Corporation, St. Joseph, MI, USA).

For bulk mineral composition measurements, the samples were mixed with ethanol, mounted on glass slides and measured on a Bruker D8 Discover diffractometer (Bruker AXS Corporation, Karlsruhe, Germany). The diffracted beam was measured with a scintillation detector with a counting time of 20 s for each step of 0.02° 2θ. The quantitative phase analysis was performed by Rietveld refinement, with customized clay mineral structure models [57].

2.2.3. FE-SEM

Shale sections of 1 cm × 1 cm in area were Ar-ion milled to create an ultra-smooth surface and then coated with carbon. Each carbon-coated section was inspected using an FEI Helios NanoLab™ 650 FE-SEM (Thermo Fisher Scientific, Waltham, MA, USA). The FE-SEM images of the shale sample surface, with a resolution of 2.5 nm at 2 kV accelerating voltage and a working distance of 4 mm, were collected.

2.2.4. LPGA

LPGA (CO_2 and N_2) experiments were performed on Micromeritics ASAP-2460 surface area analyzers (Micromeritics, Atlanta, GA, USA). The OM and shale samples were degassed at about 110 °C for about 12 h before the LPGA experiments. The parameters were set at 0 °C and −196 °C. The relative pressure (P/P_0) for N_2 and CO_2 adsorption ranged from 0.0001 to 0.995 and from 0.0001 to 0.03, respectively. The adsorption isotherms were generated and the surface areas, pore volumes, and pore distributions were calculated [13]. Micropore volumes and surface areas were calculated using a Density Functional Theory (DFT) model based on CO_2 physisorption data [37]. The Barrett-Joyner-Halenda (BJH) volumes were calculated using low-pressure N_2 adsorption data [33–37].

3. Results

3.1. Organic Geochemistry and Mineralogy

The mineral compositions and geochemical results of the W-L shale samples are listed in Table 1. The TOC content ranged from 1.5 wt% to 4.2 wt%, with an average content of 2.75 wt%. The equivalent vitrinite reflectance values, converted from bitumen reflectance, were in the range of 2.35–2.60%, suggesting a mature dry gas generation. The dominant mineral compositions in the shale were quartz and clay, with average proportions of 40.8% (35.6–51.1%) and 38.9% (38.3–41.1%), respectively. The average values of feldspar, carbonate minerals (calcite and dolomite) and pyrite were 9.6%, 4.8%, 2.7%, respectively.

Table 1. Geochemical characteristics and mineral compositions of Wufeng-Longmaxi (W-L) shale samples.

Sample ID	Depth (m)	TOC (%)	Ro (%)	Quartz (%)	Feldspar (%)	Calcite (%)	Dolomite (%)	Clay mineral (%)	Pyrite (%)
J4-1	2282.5	1.5	2.49	35.6	10.5	7.1	1.3	41.1	3.1
J4-2	2354.6	2.1	2.52	37.3	13.6	5.4	1.2	37.6	2.3
J4-3	2368.1	2.7	2.53	41.2	8.8	3.2	0.5	38.3	3.3
J4-4	2386.4	3.2	2.55	39.2	11.2	5.1	1.3	38.4	3.2
J4-5	2408.4	4.2	2.6	51.1	7.1	0.1	0.1	38.5	2.1

3.2. FE-SEM Imaging

FE-SEM images were obtained to study the pore characteristics of the W-L shale samples (Figure 1). According to the classification of Loucks et al. (2012) [2], OM pores, interparticle pores (interP pores) and intraparticle pores (intraP pores) were observed. OM pores with various pore shapes and sizes were identified (Figure 1a–c). OM pores with elliptical bubbles and irregular polygon shapes were heterogeneously distributed (Figure 1a–b). Within large OM particles, the pores extended toward each other, forming complex and connected pore networks in three dimensions (Figure 1b). Pores between the illite interlayers were also filled with OM, which formed complex pore networks (Figure 1c). These observations were consistent with the results of the W-L Formation in the Sichuan Basin [14,31,32,35,36]. InterP pores are primarily formed between quartz, clay minerals, and calcite grains with, mainly, slit shape (Figure 1 d–f). Such pores are commonly filled with OM, pyrite framboids, and clay minerals (Figure 1 g–h). IntraP pores are also observed within quartz (Figure 1d), and calcite grain (Figure 1i). Pyrite framboids with inter crystal pores filled with OM (Figure 1g) were commonly observed in the samples.

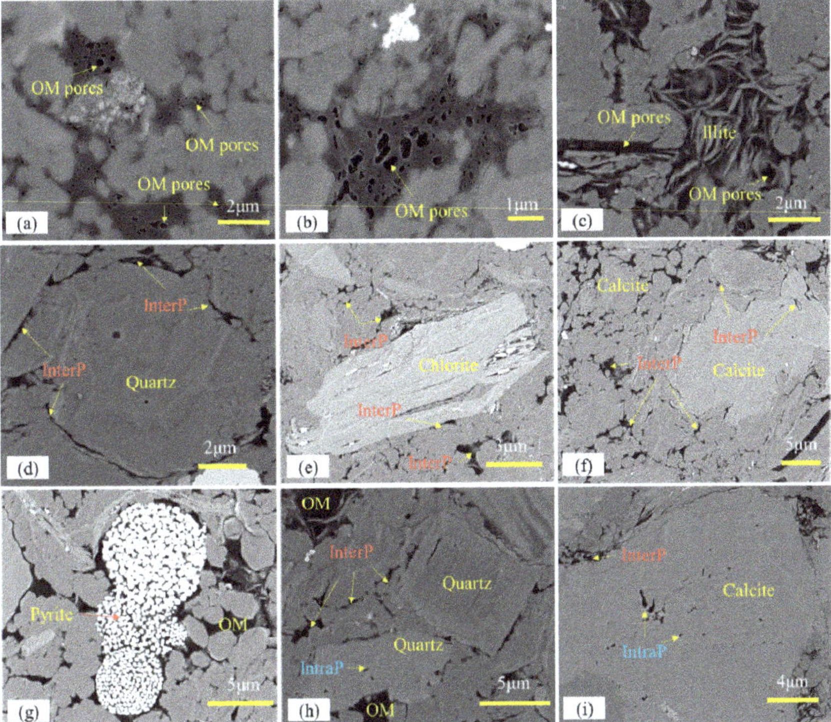

Figure 1. Example of pores in W-L shale sample J4-5. (**a**)–(**b**) Large organic matter (OM) particles with elliptical bubbles and irregular polygon shapes; (**c**) complex OM pores within the illite interlayers; (**d**)–(**f**) interP pores between quartz, clay minerals, and calcite grain rims; (**g**)–(**h**) quartz and pyrite framboids with inter crystal pores filled with OM and with OM pores; (**i**) intraP pores observed within calcite grain.

3.3. *LPGA (CO_2 and N_2)*

3.3.1. Low-Pressure CO_2 Adsorption

The low-pressure CO_2 adsorption isotherms of the isolated OM and bulk shale samples are shown in Figure 2. The OM samples had slightly higher adsorption volumes than the related bulk shale samples, suggesting the formation of micropores in OM matter. Isotherms with similar shapes were observed in the isolated OM samples (Figure 2a). For the bulk shale samples, the J4-1 shale with the lowest TOC content had the lowest CO_2 adsorption volume, while sample J4-5 with the higher OM richness had the highest CO_2 adsorption volume (Figure 2b).

The pore volumes and surface areas calculated by DFT model are listed in Table 2. The DFT surface area of isolated OM ranges from 30.98 to 34.85 m^2/g, with an average of 32.58 m^2/g, which is nearly two times larger than the related shale samples with a mean value of 9.32 m^2/g (12.68–28.95 m^2/g). The DFT pore volume of the isolated OM samples varies from 12.6 to 13.2 $cm^{3\cdot}/100g$, which is greater than that of the corresponding bulk shales (0.32–1.12 $cm^3/100g$). In addition, when the TOC content increased, the DFT pore volume of all bulk shale samples increased. The pore volumes of the OM samples were normalized to bulk shale weight by multiplying the TOC values of the bulk shale. The normalized pore volumes of OM are in the range 0.19–0.55 $cm^3/100g$ shale (Table 2).

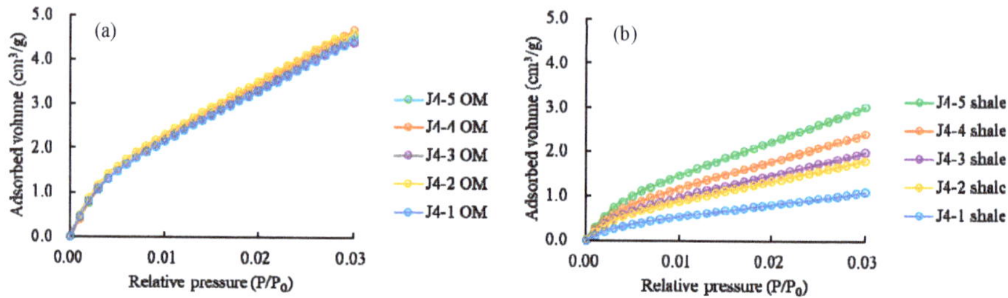

Figure 2. Comparison of low-pressure CO_2 adsorption isotherms of (**a**) isolated OM samples and (**b**) bulk shale samples.

Table 2. Pore structure parameters of isolated OM and shale from low-pressure CO_2 adsorption data.

Sample ID	Bulk Shales		Isolated OM		
	DFT Surface Area (m^2/g)	DFT Pore Volume (cm^3/100g shale)	DFT Surface Area (m^2/g)	DFT Pore Volume (cm^3/100g)	DFT Pore Volume (cm^3/100g shale)
J4-1	12.68	0.32	32.61	13.0	0.19
J4-2	15.92	0.46	31.13	12.8	0.26
J4-3	16.88	0.51	33.34	13.6	0.36
J4-4	24.12	0.95	30.98	12.6	0.40
J4-5	28.95	1.12	34.85	13.2	0.55

The PSDs of isolated OM and bulk shale samples obtained from CO_2 adsorption data are shown in Figure 3. Pores with diameters smaller than 0.7 nm are the primary proportions of pore volume in both isolated OM and bulk shales (Figure 3). The shapes of PSD curves of both isolated OM and bulk shale samples are similar. These results are consistent with previous studies of W-L shale in the Sichuan Basin [35–37].

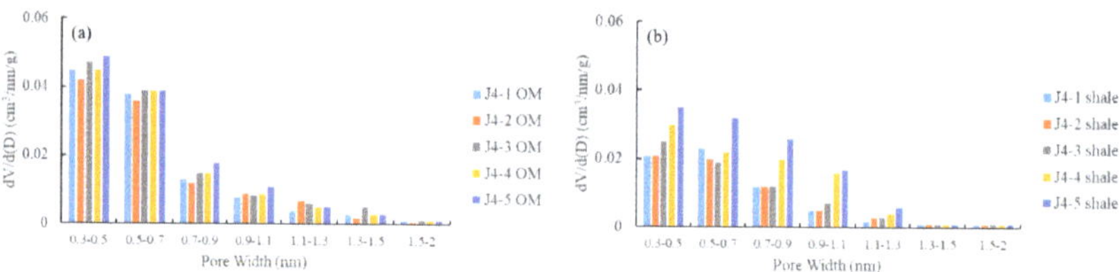

Figure 3. Pore volumes (**a**)–(**b**) distribution derived from low-pressure CO_2 adsorption isotherms for (**a**) OM and (**b**) bulk shale samples.

3.3.2. Low-Pressure N_2 Physisorption

The low-pressure N_2 physisorption isotherms of isolated OM and bulk shale samples are illustrated in Figure 4. Type IV isotherms can be identified [55]. These isotherms represent both mesopores and macropores in isolated OM and bulk shales. Isotherms of samples J4-3, J4-4 and J4-5 (TOC > 2.5 wt%) show adsorption at a low relative pressure, indicating the existence of micropores [13,40,58]. These characteristics are consistent with the low-pressure CO_2 adsorption data. OM samples have more N_2 adsorption volume than related bulk shale samples, indicating that OM is more porous than other fractions of the studied shale samples.

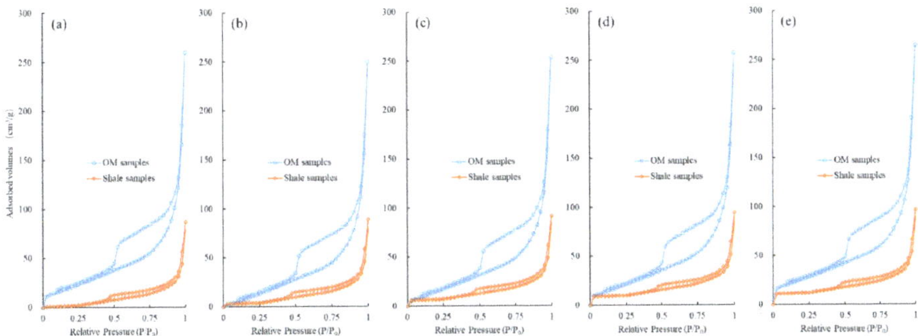

Figure 4. Comparison of low-pressure N_2 adsorption isotherms of isolated OM and bulk shale samples. From left to right, TOC content of the samples increased; (**a**) J4-1 sample; (**b**) J4-2 sample; (**c**) J4-3 sample; (**d**) J4-4 sample; (**e**) J4-5 sample.

The pore structure parameters obtained from N_2 adsorption isotherms are presented in Table 3. The average Brunauer-Emmett-Teller (BET) surface area of the isolated OM in this study was 79.16 m^2/g (70.22–86.23 m^2/g), which is nearly three times larger than that of the bulk shale, 28.86 m^2/g (27.21–32.34 m^2/g). The Barrett-Joyner-Halenda (BJH) pore volume of isolated OM varied from 35.43 to 40.80 cm^3/g, with a mean value of 37.64 cm^3/g. This was about three times greater than that of the bulk shale samples, which ranged from 8.2 to 14.27 cm^3/g, with a mean value of 10.92 cm^3/g. The pore volumes of the OM samples were normalized to bulk shale weight, and the normalized BJH pore volumes of OM ranged from 0.53–1.46 cm^3/100g shale (Table 3).

Table 3. Pore structure parameters of isolated OM and shale from low-pressure N_2 adsorption data.

Sample ID	Bulk Shales			Isolated OM			
	BET Surface Area (m^2/g)	BJH Pore volume (cm^3/100 g)	Average Pore Size (nm)	BET Surface Area (m^2/g)	BJH Pore Volume (cm^3/100 g)	BJH Pore Volume (cm^3/100 g shale)	Average Pore Size (nm)
J4-1	27.21	8.2	9.16	86.23	35.43	0.53	17.78
J4-2	28.20	12.29	11.44	83.28	40.80	0.86	19.72
J4-3	29.33	10.35	9.59	80.21	38.13	1.03	19.75
J4-4	27.24	9.48	11.48	75.87	39.17	1.25	22.02
J4-5	32.34	11.27	14.15	70.22	34.67	1.46	21.72

The PSDs calculated using low-pressure N_2 adsorption data of isolated OM and bulk shales are shown in Figure 5. The PSDs calculated from the desorption branch of isotherms created fake peaks at around 4 nm, due to the tensile strength effect [58]. Therefore, the adsorption branches were used to calculate PSD. The PSD spectra of the OM and bulk shale samples showed similar trends with an obvious unimodal nature of the curves within the pore size of about 3–5nm. This result is consistent with other results of Longmaxi shale from the Sichuan Basin [33–37,52–54].

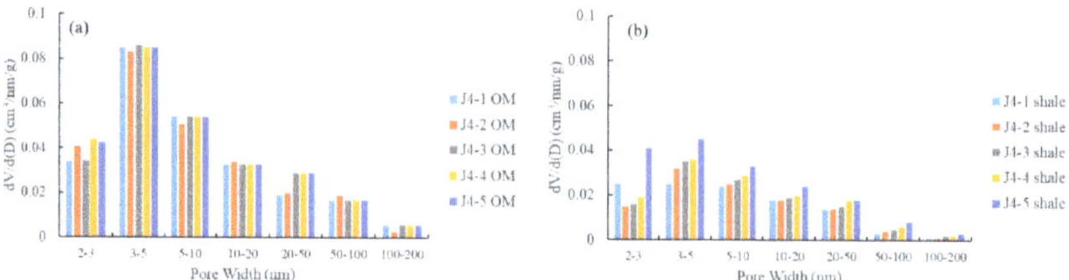

Figure 5. Pore volumes (**a**)–(**b**) distribution derived from low-pressure CO_2 adsorption isotherms for (**a**) OM and (**b**) bulk shale samples.

4. Discussion

4.1. Comparison of PSDs between Isolated OM and Bulk Shales

The PSDs of isolated OM and bulk shale samples were compared to understand the different impacts of OM and inorganic minerals on pore characteristics in shale. The pore volumes of the OM samples were normalized to bulk shale weight by multiplying the TOC values of the bulk shale (Figure 6). The curves of dV/dlog(D) represented the derivative of the y-axis value, which was amplified for larger pores to better illustrate the characteristics [3]. The dV/dlog(D) spectra of both isolated OM and bulk shale showed multimodal characteristics in the pore sizes of 0.3–0.8 nm, 3–5 nm and 70–100 nm (Figure 6).

The PSDs of the isolated OM samples showed similar shapes and trends (Figure 6a–e), indicating that OM pores may have similar shapes and pore widths. The homogeneities of OM type and thermal maturity in the W-L shale samples in this work evidence the similar PSD curves of the isolated OM. In addition, the similar shapes of the PSD curves may also support the application of OM isolation from bulk shale for the characterization of OM pores [39,40,45].

The PSD curves of the bulk shale samples present slightly different trends according to diverse TOC contents. When the TOC content increased, the pore volumes of the bulk shale samples all increased (Figure 6). The divergences of the PSD curves between the isolated OM and their corresponding bulk shale samples may provide information for the different effects of OM and minerals on pore development in the W-L gas shales [40]. The smaller the divergence, the greater the proportion of OM pore to bulk shale. The larger the divergence, the greater the proportion of inorganic mineral host pore to bulk shale. As can be seen in Figure 6, with the TOC content of the bulk shale increasing, the divergences of the PSD curves of the isolated OM samples and their bulk shale samples gradually decreased (Figure 6). These results suggest that inorganic minerals host abundant pores and that mineral compositions may contribute mainly to the macropores in gas shale.

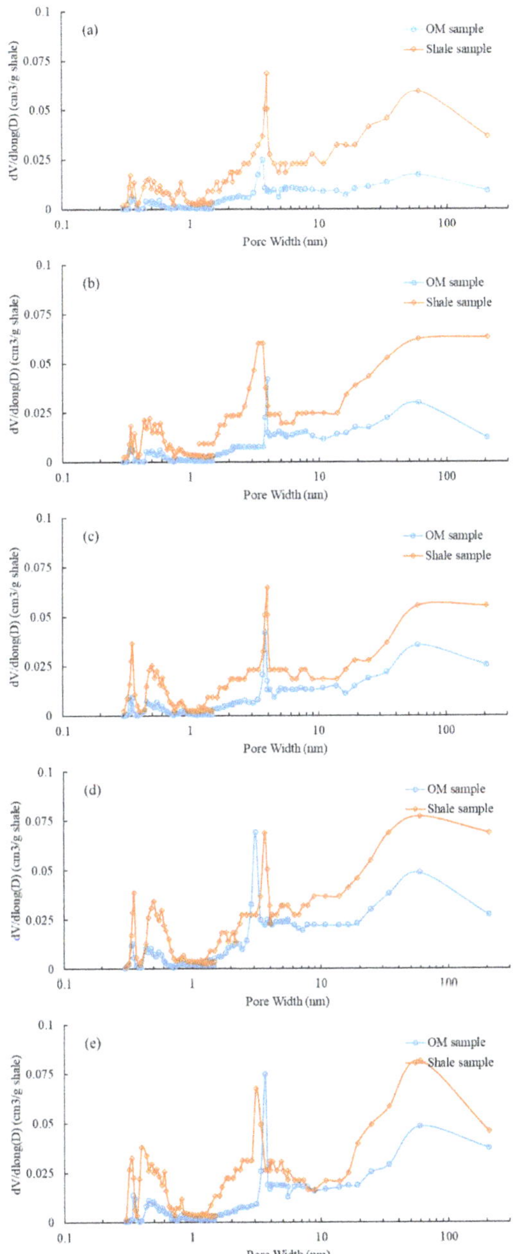

Figure 6. Comparison of pore size distribution between isolated OM and its corresponding bulk shale sample; (**a**) J4-1 sample; (**b**) J4-2 sample; (**c**) J4-3 sample; (**d**) J4-4 sample; (**e**) J4-5 sample.

4.2. Contributions of OM Pore Volumes

The pore volumes of micropore (<2 nm), mesopore (2–50 nm) and macropore (>50 nm) calculated from LPGA data are shown in Table 4 and Figure 7. The pore volumes of the OM samples are much larger than those in the corresponding shale samples. The micropore volumes were calculated based on CO_2 adsorption data. Mesopore and macropore volumes were obtained from N_2 adsorption data. For the bulk shale samples, the micropore, mesopore, and macropore volumes varied in the ranges of 0.32–1.12, 3.7–5.38 and 4.5–7.71 $cm^3/100$ g, respectively. For the OM samples, the pore volumes of the micropores varied from 12.6 to 13.6 $cm^3/100$ g. The pore volumes of the mesopores and macropores were in the ranges 11.11–19.05 and 19.6–22.23 $cm^3/100$ g, respectively (Figure 7). The gas (CO_2 and N_2) volumes, pore volumes, surface areas, PSD lines and porosities of the organic matter samples were all larger than those of the shale samples, which indicates that organic matter (OM) is more porous than other fractions of the studied shale samples. The results are consistent with the previous study of the Longmaxi shale by Tian et al. (2013), who proposed that OM had more micropores and fine mesopores than clay minerals in Longmaxi shale [23].

Table 4. Pore volumes of isolated OM and shale obtained from low pressure gas (CO_2 and N_2) adsorption (LPGA) data.

Sample ID	Bulk Shales (cm^3/100 g)				Isolated OM (cm^3/100 g)			
	Micropores	Mesopores	Macropores	Total	Micropores	Mesopores	Macropores	Total
J4-1	0.32	3.70	4.50	8.52	13.04	15.83	19.60	48.43
J4-2	0.46	4.58	7.71	12.75	12.80	16.57	24.23	53.60
J4-3	0.51	4.14	6.21	10.86	13.62	16.17	21.96	51.73
J4-4	0.95	3.98	5.50	10.43	12.61	19.05	20.12	51.77
J4-5	1.12	5.38	5.89	12.39	13.21	11.11	23.56	47.87

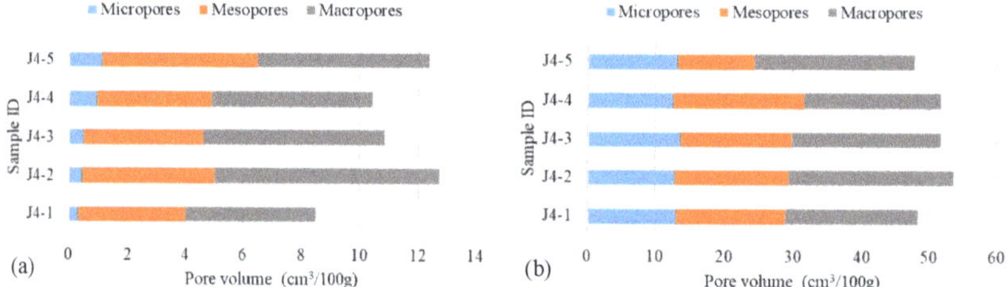

Figure 7. Comparison pore proportions in (**a**) bulk shales and(**b**) isolated OM samples.

OM porosity is primarily controlled by organic richness and thermal maturity [13,16,59–61]. Previous studies reported that there were positive correlations between porosity and TOC content, suggesting that the TOC content is the primary control of the organic porosity [15,16,19,23,33,34,36–38]. When equivalent vitrinite reflectance values are in the range of 2.0–2.5%, the OM pores in shale commonly have large diameters, up to tens of hundreds of nanometers in FE-SEM images [16,49,59–62]. The eq. Ro values of the W-L shale samples in this work were in the range 2.35–2.60%, during the generation of shale gas, shale porosity significantly increased, which was supported by the FE-SEM images showing large OM pores in the samples (Figure 1a–b). In addition, the PSD spectra of the isolated OM samples showed the obvious unimodal nature of the curves within a pore size of about 0.3–0.8, 3–5 and 70–100 nm (Figure 6). The positive correlations were slight in many gas shales, which was probably due to the minor amount of OM in the shale samples [35–37].

The pore structure in gas shales may also be controlled by inorganic minerals. Numerous studies have proposed the importance of clay minerals for shale porosity [13,17,33,63] and pore structure [31–33]. In this study, the framework of clay flakes exerts an important influence on the formation of pores in clay minerals (Figure 1e–f). These pores are mostly developed in the pressure shadow of hard grains with compaction resistance [61]. Quartz can form rigid frameworks to prevent pores from collapsing so that the primary porosity is well preserved [2]. Highly-mature shale is commonly buried deep. Therefore, OM pores may be lost during compaction due to the soft and ductile nature of the OM [2,16,17]. FE-SEM images showed that the OM pores surrounded by rigid framework minerals were more easily preserved (Figure 1d–i). IntraP pores within calcite grain suggested that dissolution of carbonate minerals is a vital mechanism for the development of intraparticle pores in shale [2]. The pores related to mineral compositions resulted in more complicated and heterogeneous pore size characteristics of bulk shales [2,64].

The contributions of OM and inorganic composition to the total pore volumes were calculated by converting all pore volume per OM weight to per shale weight ($cm^3/100$ g OM to $cm^3/100$ g shale). The results are shown in Table 5 and Figure 8. The contributions of OM to the micropore volume ranged from 49.5% to 72.0%, with an average of 58.42%. The contributions of OM to mesopore volume were between 6.4% and 18.5%, and averaged 10.34%. The contributions of OM to macropore volume were in the range 6.5–16.8%, with an average of 10.72%. In total, OM pores accounted for 8.5–19.2% of the total volume of the W-L shale (Figure 8 a–c).

Table 5. Contributions of OM and inorganic minerals to the total pore volume of the W-L shales.

Sample ID	Contributions to the Total Volume of Shales							
	OM				Inorganic Compositions			
	Micropores (%)	Mesopores (%)	Macropores (%)	Total (%)	Micropores (%)	Mesopores (%)	Macropores (%)	Total (%)
J4-1	60.9	6.4	6.5	8.5	39.1	93.6	93.5	91.5
J4-2	58.4	7.6	6.6	8.8	41.6	92.4	93.4	91.2
J4-3	72.0	10.5	9.5	12.9	28.0	89.5	90.5	87.1
J4-4	51.3	18.5	14.2	19.2	48.7	81.5	85.8	80.8
J4-5	49.5	8.7	16.8	16.2	50.5	91.3	83.2	83.8

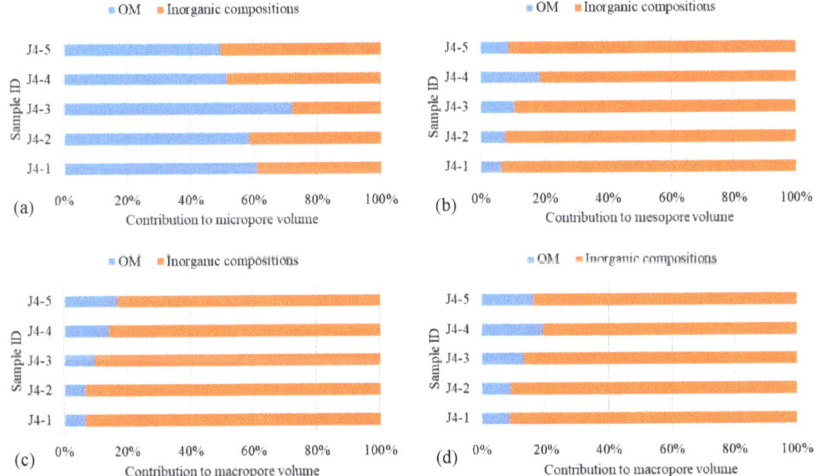

Figure 8. Contributions of OM and inorganic compositions to (**a**) micropore volume, (**b**) mesopore volume, (**c**) macropore volume and (**d**) and total pore volume of W-L shale samples.

The contributions of OM pores to the total pore volumes (Figure 8d) were lower than those of the inorganic composition. This result is consistent with a previous study of the contribution of OM porosity to Longmaxi shale (31.3–62.6%) [23], but inconsistent with the corresponding contribution value of Barnett shale (95.2%). Tian et al. (2013) [23] calculated the contribution of OM pores in Longmaxi shale using the correlation of the TOC content versus porosity. In their calculation they assumed that the contribution of inorganic mineral host pores was constant. Loucks et al. (2012) [2]

estimated the contribution of OM pores to the total pore volume of Barnett shale using FE-SEM images. However, SEM is not appropriate for pores smaller than 5 nm [17]. In addition, the spatial limitation and biased selection of the observed areas may not be able to catch representative images, resulting in the relatively high contribution of OM pores (95.2%). Therefore, this paper may provide more accurate results and new evidences of the contributions of OM and mineral compositions to the total pore volume of W-L shale.

5. Conclusions

The contributions of OM to the total pore volume of W-L shale from the Fuling gas field, Sichuan Basin were investigated through FE-SEM observation, OM isolation, and LPGA (CO_2 and N_2) analysis. Based on our results, the following conclusions can be drawn.

(1) Abundant OM pores and inorganic pores with various pore shapes and widths can be observed in FE-SEM images in the W-L shales.

(2) The LPGA adsorption volume, pore volume, and pore surface area of the OM samples were all larger than those of the bulk shale samples, indicating that OM is more porous than inorganic compositions in W-L shale.

(3) The average contribution of organic matter to the volume of micropore, mesopore and macropore were 58.42%, 10.34% and 10.72%, respectively. The contributions of OM to the total pore volume were lower than those of the inorganic compositions, probably due to the small weight ratio of OM in shale samples (1.5 wt%–4.2 wt%).

(4) OM has more micropores than inorganic compositions, which makes their PSD more complicated.

Author Contributions: Z.L. (Zhuo Li) and Z.J. designed the project; Z.L. (Zhikai Liang) performed the experiments and analyzed the data; Z.L. (Zhuo Li) wrote this paper and Z.J. corrected it; Z.L. (Zhikai Liang) and H.Y. modified the formats.

Funding: This study was supported by the by the National Natural Science Foundation of China (No. 41502123 and No. U1562215) and the National Major Project of China (No. 2017ZX05035-002 and No. 201605034-001).

Acknowledgments: We would like to thank the Jianghan Oil Field Company Branch, SINOPEC for providing the shale samples. We would also like to thank the Research Institute of Petroleum Exploration and Development, Petrochina for providing the laboratory for the OM isolation. We are grateful to the editors and reviewers for their constructive comments and suggestions.

Conflicts of Interest: The authors declare no conflicts of interest.

References

1. Ross, D.; Bustin, R. The importance of shale composition and pore structure upon gas storage potential of shale gas reservoirs. *Mar. Pet. Geol.* **2009**, *26*, 916–927. [CrossRef]
2. Loucks, R.; Reed, R.; Ruppel, S.; Hammes, U. Spectrum of pore types and networks in mudrocks and a descriptive classification for matrix-related mudrock pores. *AAPG Bull.* **2012**, *96*, 1071–1098. [CrossRef]
3. Clarkson, C.R.; Solano, N.; Bustin, R.M.; Bustin, A.M.M.; Chalmers, G.R.L.; He, L.; Melnichenko, Y.B.; Radlinski, A.P.; Blach, T.P. Pore structure characterization of North American shale gas reservoirs: Using USANS/SANS, gas adsorption, and mercury intrusion. *Fuel.* **2013**, *103*, 606–616. [CrossRef]
4. Sing, K.S.; Everett, D.H.; Haul, R.A.W.; Moscou, L.; Pierotti, R.A.; Rouquerol, J.; Siemieniewsha, T. Reporting physisorption data for gas/solid systems with special reference to the determination of surface area and porosity. *Pure Appl.Chem.* **1985**, *57*, 603–619. [CrossRef]
5. Wu, K.L.; Chen, Z.X.; Li, X.F. Real gas transport through nanopores of varying cross-section type and shape in shale gas reservoirs. *Chem. Eng. J.* **2015**, *281*, 813–825. [CrossRef]
6. Wu, K.L.; Chen, Z.X.; Li, X.F.; Xu, J.Z.; Li, J.; Wang, K. Flow behavior of gas confined in nanoporous shale at high pressure: Real gas effect. *Fuel* **2017**, *205*, 173–183. [CrossRef]
7. Collell, J.; Galliero, G.; Vermorel, R.; Ungerer, P.; Yiannourakou, M.; Montel, F. Transport of multicomponent hydrocarbon mixtures in shale organic matter by molecular simulations. *J. Phys. Chem. C* **2015**, *119*, 22587–22595. [CrossRef]

8. Lee, T.; Bocquet, L.; Coasne, B. Activated desorption at heterogeneous interfaces and long-time kinetics of hydrocarbon recovery from nanoporous media. *Nat. Commun.* **2016**, *7*, 11890. [CrossRef] [PubMed]
9. Javadpour, F.; McClure, M.; Naraghi, M.E. Slip-corrected liquid permeability and its effect on hydraulic fracturing and fluid loss in shale. *Fuel* **2015**, *160*, 549–559. [CrossRef]
10. Naraghi, M.E.; Javadpour, F. A stochastic permeability model for the shale-gas systems. *Int. J. Coal Geol.* **2015**, *140*, 111–124. [CrossRef]
11. Naraghi, M.E.; Javadpour, F.; Ko, L.T. An object-based shale permeability model: Non-Darcy gas flow, sorption, and surface diffusion effects. *Transp. Porous Med.* **2018**, *125*, 23–39. [CrossRef]
12. Berthonneau, J.; Obliger, A.; Valdenaire, P.L.; Grauby, O.; Ferry, D.; Chaudanson, D. Mesoscale structure, mechanics, and transport properties of source rocks' organic pore networks. *Proc. Nat. Acad. Sci.* **2018**, *115*, 12365–12370. [CrossRef]
13. Loucks, R.G.; Reed, R.M.; Ruppel, S.C.; Jarvie, D.M. Morphology, genesis, and distribution of nanometer-scale pores in siliceous mudstones of the Mississippian Barnett Shale. *J. Sediment. Res.* **2009**, *79*, 848–861. [CrossRef]
14. Wang, P.; Jiang, Z.; Ji, W.; Zhang, C.; Yuan, Y.; Chen, L. Heterogeneity of intergranular, intraparticle and organic pores in Longmaxi shale in Sichuan basin, south China: Evidence from SEM digital images and fractal and multifractal geometries. *Mar. Pet. Geol.* **2016**, *72*, 122–138. [CrossRef]
15. Milliken, K.L.; Esch, W.L.; Reed, R.M.; Zhang, T. Grain assemblages and strong diagenetic overprinting in siliceous mudrocks, Barnett Shale (Mississippian), Fort Worth Basin, Texas, U.S.A. *AAPG Bull.* **2012**, *96*, 1553–1578. [CrossRef]
16. Milliken, K.L.; Rudnicki, M.; Awwiller, D.N.; Zhang, T. Organic matter-hosted pore system, Marcellus Formation (Devonian), Pennsylvania. *AAPG Bull.* **2013**, *97*, 177–200. [CrossRef]
17. Chalmers, G.R.; Bustin, R.M.; Power, I.M. Characterization of gas shale pore systems by porosimetry, pycnometry, surface area, and field emission scanning electron microscopy/transmission electron microscopy image analyses: Examples from the Barnett, Woodford, Haynesville, Marcellus, and Doig units. *AAPG Bull.* **2012**, *96*, 1099–1119.
18. Jiang, F.J.; Chen, J.; Xu, Z.; Wang, Z.F.; Hu, T.; Chen, D. Organic matter pore characterization in lacustrine shales with variable maturity using nanometer-scale resolution X-ray computed tomography. *Energ. Fuel.* **2017**, *31*, 2669–2680. [CrossRef]
19. Mastalerz, M.; He, L.; Melnichenko, Y.B.; Rupp, J.A. Porosity of coal and shale: Insights from gas adsorption and SANS/USANS techniques. *Energ. Fuel.* **2012**, *26*, 5109–5120. [CrossRef]
20. Sun, M.; Yu, B.; Hu, Q.; Zhang, Y.; Li, B.; Yang, R.; Melnichenkoe, Y.B.; Cheng, G. Pore characteristics of Longmaxi shale gas reservoir in the Northwest of Guizhou, China: Investigations using small-angle neutron scattering (SANS), helium pycnometry, and gas sorption isotherm. *Int. J. Coal Geol.* **2017**, *171*, 61–68. [CrossRef]
21. Wang, G.; Ju, Y.; Yan, Z.; Li, Q. Pore structure characteristics of coal-bearing shale using fluid invasion methods: A case study in the Huainan–Huaibei Coalfield in China. *Mar. Pet. Geol.* **2015**, *62*, 1–13. [CrossRef]
22. Labani, M.M.; Rezaee, R.; Saeedi, A.; Al Hinai, A. Evaluation of pore size spectrum of gas shale reservoirs using low pressure nitrogen adsorption, gas expansion and mercury porosimetry: A case study from the Perth and Canning Basins, Western Australia. *J. Petrol. Sci. Eng.* **2013**, *112*, 7–16. [CrossRef]
23. Tian, H.; Pan, L.; Xiao, X.M.; Wilkins, R.W.T.; Meng, Z.P.; Huang, B.J. A preliminary study on the pore characterization of Lower Silurian black shales in the Chuandong Thrust Fold Belt, southwestern China using low pressure N_2 adsorption and FE-SEM methods. *Mar. Pet. Geol.* **2013**, *48*, 8–19. [CrossRef]
24. Sigal, R.F. Pore-size distributions for organic-shale-reservoir rocks from nuclear magnetic resonance spectra combined with adsorption measurements. *SPE J.* **2015**, *20*, 1–7. [CrossRef]
25. Fleury, M.; Romero-Sarmiento, M. Characterization of shales using T_1–T_2 NMR maps. *J. Petrol. Sci. Eng.* **2016**, *137*, 55–62. [CrossRef]
26. Meng, M.; Ge, H.; Ji, W.; Shen, Y.; Shuai, S. Monitor the process of shale spontaneous imbibition in co-current and counter-current displacing gas by using low field nuclear magnetic resonance method. *J. Nat. Gas. Sci. Eng.* **2015**, *27*, 336–345. [CrossRef]
27. Zhang, P.; Lu, S.; Li, J.; Chen, C.; Xue, H.; Zhang, J. Petrophysical characterization of oil-bearing shales by low-field nuclear magnetic resonance (NMR). *Mar. Petrol. Geol.* **2018**, *9*, 775–785. [CrossRef]

28. Li, A.; Ding, W.; Jiu, K.; Wang, Z.; Wang, R.; He, J. Investigation of the pore structures and fractal characteristics of marine shale reservoirs using NMR experiments and image analyses: A case study of the Lower Cambrian Niutitang Formation in northern Guizhou Province, South China. *Mar. Pet. Geol.* **2018**, *89*, 530–540. [CrossRef]
29. Zhou, L.; Kang, Z. Fractal characterization of pores in shales using NMR: A case study from the Lower Cambrian Niutitang Formation in the Middle Yangtze Platform, Southwest China. *J. Nat. Gas. Sci. Eng.* **2016**, *35*, 860–872. [CrossRef]
30. Wang, Y.; Pu, J.; Wang, L.H.; Wang, J.Q.; Jiang, Z.; Song, F.Y. Characterization of typical 3D pore networks of Jiulaodong formation shale using nano-transmission X-ray microscopy. *Fuel* **2016**, *170*, 84–91. [CrossRef]
31. Tang, X.L.; Jiang, Z.X.; Jiang, S.; Li, Z. Heterogeneous nanoporosity of the Silurian Longmaxi Formation shale gas reservoir in the Sichuan Basin using the QEMSCAN, FIB-SEM, and nano-CT methods. *Mar. Petrol. Geol.* **2016**, *78*, 99–109. [CrossRef]
32. Wang, P.F.; Jiang, Z.X.; Chen, L.; Yin, L.S.; Li, Z.; Zhang, C.; Tang, X.L.; Wang, G.Z. Pore structure characterization for the Longmaxi and Niutitang shales in the Upper Yangtze Platform, South China: Evidence from focused ion beam-He ion microscopy, nano-computerized tomography and gas adsorption analysis. *Mar. Petrol. Geol.* **2016**, *77*, 1323–1337. [CrossRef]
33. Yang, F.; Ning, Z.F.; Wang, Q.; Zhang, R.; Krooss, B.M. Pore structure characteristics of lower Silurian shales in the southern Sichuan Basin, China: Insights to pore development and gas storage mechanism. *Int. J. Coal Geol.* **2016**, *156*, 12–24. [CrossRef]
34. Shao, X.; Pang, X.; Li, Q.; Wang, P.; Di, C.; Shen, W. Pore structure and fractal characteristics of organic-rich shales: A case study of the lower Silurian Longmaxi shales in the Sichuan basin, SW china. *Mar. Pet. Geol.* **2017**, *80*, 192–202. [CrossRef]
35. Jiang, Z.X.; Tang, X.H.; Cheng, L.J.; Li, Z.; Zhang, Y.Y.; Bai, Y.Q. Characterization and origin of the Silurian Wufeng-Longmaxi Formation shale multiscale heterogeneity in southeastern Sichuan Basin, China. *Interpretation* **2015**, *3*, 61–74. [CrossRef]
36. Yang, R.; He, S.; Yi, J.; Hu, Q.H. Nano-scale pore structure and fractal dimension of organic-rich Wufeng-Longmaxi shale from Jiaoshiba area, Sichuan Basin: Investigations using FE-SEM, gas adsorption and helium pycnometry. *Mar. Petrol. Geol.* **2016**, *70*, 27–45. [CrossRef]
37. Chen, L.; Jiang, Z.X.; Liu, K.Y.; Tan, J.; Gao, F.L.; Wang, P.F. Pore structure characterization for organic-rich Lower Silurian shale in the Upper Yangtze Platform, South China: A possible mechanism for pore development. *J. Nat. Gas. Sci. Eng.* **2017**, *46*, 1–15. [CrossRef]
38. Yang, F.; Ning, Z.; Wang, Q.; Liu, H.Q. Pore structure of Cambrian shales from the Sichuan Basin in China and implications to gas storage. *Mar. Petrol. Geol.* **2016**, *70*, 14–26. [CrossRef]
39. Rexer, T.F.; Mathia, E.J.; Aplin, A.C.; Thomas, K.M. High-pressure methane adsorption and characterization of pores in Posidonia shales and isolated kerogens. *Energy Fuels* **2014**, *28*, 2886–2901. [CrossRef]
40. Ji, W.M.; Song, Y.; Rui, Z.; Meng, M.M.; Huang, H.X. Pore characterization of isolated organic matter from high matured gas shale reservoir. *Int. J. Coal. Geol.* **2017**, *174*, 31–40. [CrossRef]
41. Han, H.; Pang, P.; Li, Z.L.; Shi, P.T.; Guo, C.; Liu, Y.; Chen, S.J.; Lu, J.G.; Gao, Y. Controls of organic and inorganic compositions on pore structure of lacustrine shales of Chang 7 member from Triassic Yanchang Formation in the Ordos Basin, China. *Mar. Petrol. Geol.* **2019**, *100*, 270–284. [CrossRef]
42. Ji, W.M.; Song, Y.; Jiang, Z.X.; Chen, L.; Li, Z.; Yang, X.; Meng, M.M. Estimation of marine shale methane adsorption capacity based on experimental investigations of Lower Silurian Longmaxi formation in the Upper Yangtze Platform, south China. *Mar. Pet. Geol.* **2015**, *68*, 94–106. [CrossRef]
43. Jarvie, D.M.; Hill, R.J.; Ruble, T.E.; Pollastro, R.M. Unconventional shale-gas systems: The Mississippian Barnett Shale of northcentral Texas as one model for thermogenic shale-gas assessment. *AAPG Bull.* **2007**, *91*, 475–499. [CrossRef]
44. Zhang, T.W.; Ellis, G.S.; Ruppel, S.C.; Milliken, K.; Yang, R.S. Effect of organicmatter type and thermal maturity on methane adsorption in shale-gas systems. *Org. Geochem.* **2012**, *47*, 120–131. [CrossRef]
45. Gasparik, M.; Bertier, P.; Gensterblum, Y.; Ghanizadeh, A.; Krooss, B.M.; Littke, R. Geological controls on the methane storage capacity in organic-rich shales. *Int. J. Coal. Geol.* **2014**, *123*, 34–51. [CrossRef]
46. Jiang, F.J.; Chen, D.; Wang, Z.F.; Xu, Z.Y.; Chen, J.; Liu, L. Pore characteristic analysis of a lacustrine shale: A case study in the Ordos Basin, NW China. *Mar. Petrol. Geol.* **2016**, *73*, 554–571. [CrossRef]

47. Curtis, M.E.; Ambrose, R.J.; Sondergeld, C.H. Structural Characterization of Gas Shales on the Micro-and Nano-Scales. In Proceedings of the Canadian Unconventional Resources and International Petroleum Conference, Calgary, Alberta, Canada, 19–21 October 2010.
48. Sondergeld, C.H.; Ambrose, R.J.; Rai, C.S.; Moncrieff, J. Microstructural studies of gas shales. *SPE J.* **2010**, *2*, 23–25.
49. Curtis, M.E.; Cardott, B.J.; Sondergeld, C.H.; Rai, C.S. Development of organic porosity in the Woodford shale with increasing thermal maturity. *Int. J. Coal. Geol.* **2012**, *103*, 26–31. [CrossRef]
50. Ma, Y.; Zhong, N.N.; Li, D.H.; Pan, Z.J.; Liu, K.Y. Organic matter/clay mineral intergranular pores in the Lower Cambrian Lujiaping Shale in the north-eastern part of the upper Yangtze area, China: A possible microscopic mechanism for gas preservation. *Int. J. Coal Geol.* **2015**, *137*, 38–54. [CrossRef]
51. Chalmers, G.R.L.; Bustin, R.M. A multidisciplinary approach in determining the maceral (kerogen type) and mineralogical composition of Upper Cretaceous Eagle Ford Formation: Impact on pore development and pore size distribution. *Int. J. Coal Geol.* **2017**, *171*, 93–110. [CrossRef]
52. Sun, M.D.; Yu, B.S.; Hu, Q.H.; Yang, R.; Zhang, Y.F.; Li, B. Pore connectivity and tracer migration of typical shales in south China. *Fuel* **2017**, *203*, 32–46. [CrossRef]
53. Zhang, H.; Zhu, Y.; Wang, Y.; Kang, W.; Chen, S.B. Comparison of organic matter occurrence and organic nanopore structure within marine and terrestrial shale. *J. Nat. Gas. Sci. Eng.* **2016**, *32*, 356–363. [CrossRef]
54. Liang, C.; Jiang, Z.X.; Yang, Y.T.; Wei, X.J. Shale lithofacies and reservoir space of the Wufeng–Longmaxi formation, Sichuan Basin, China. *Pet. Explor. Dev.* **2012**, *39*, 736–743. [CrossRef]
55. Han, H.; Cao, Y.; Chen, S.; Lu, J.G.; Huang, C.X.; Zhu, H.H. Influence of particle size on gas-adsorption experiments of shales: An example from a Longmaxi Shale sample from the Sichuan Basin, China. *Fuel* **2016**, *186*, 750–757. [CrossRef]
56. Wei, M.M.; Xiong, Y.Q.; Zhang, L.; Li, J.H.; Peng, P.A. The effect of sample particle size on the determination of pore structure parameters in shales. *Int. J. Coal Geol.* **2016**, *163*, 177–185. [CrossRef]
57. Ufer, K.; Stanjek, H.; Roth, G.; Dohrmann, R.; Kleeberg, R.; Kaufhold, S. Quantitative phase analysis of bentonites by the Rietveld method. *Clays Clay Miner.* **2008**, *56*, 272–282. [CrossRef]
58. Rouquerol, J.; Avnir, D.; Fairbridge, C.W.; Everett, D.H.; Haynes, J.H.; Pernicone, N.; Ramsay, J.D.F.; Sing, K.S.W.; Unger, K.K. Recommendations for the characterization of porous solids (Technical report). *Pure Appl. Chem.* **1994**, *66*, 1739–1758. [CrossRef]
59. Mastalerz, M.; Schimmelmann, A.; Drobniak, A.; Chen, Y.Y. Porosity of Devonian and Mississippian New Albany Shale across a maturation gradient: Insights from organic petrology, gas adsorption, and mercury intrusion. *AAPG Bull.* **2013**, *97*, 1621–1643. [CrossRef]
60. Schieber, J. Common themes in the formation and preservation of intrinsic porosity in shales and mudstones: Illustrated with examples from across the Phanerozoic. *SPE J.* **2010**, *2*, 23–25.
61. Pommer, M.; Milliken, K. Pore types and pore-size distributions across thermal maturity, Eagle Ford Formation, southern Texas. *AAPG Bull.* **2015**, *99*, 1713–1744. [CrossRef]
62. Wang, G.; Ju, Y.; Han, K. Early Paleozoic shale properties and gas potential evaluation in Xiuwu basin, western lower Yangtze platform. *J. Nat. Gas. Sci. Eng.* **2015**, *22*, 489–497. [CrossRef]
63. Slatt, R.M.; O'Brien, N.R. Pore types in the Barnett and Woodford gas shales: Contribution to understanding gas storage and migration pathways in fine-grained rocks. *AAPG Bull.* **2011**, *95*, 2017–2030. [CrossRef]
64. Kuila, U.; McCarty, D.K.; Derkowski, A.; Fischer, T.B.; Topór, T.; Prasad, M. Nano-scale texture and porosity of organic matter and clay minerals in organic-rich mudrocks. *Fuel* **2014**, *135*, 359–373. [CrossRef]

© 2019 by the authors. Licensee MDPI, Basel, Switzerland. This article is an open access article distributed under the terms and conditions of the Creative Commons Attribution (CC BY) license (http://creativecommons.org/licenses/by/4.0/).

Article

Dynamic Crack Initiation Toughness of Shale under Impact Loading

Guoliang Yang [1,2], Xuguang Li [1,*], Jingjiu Bi [1] and Shuaijie Cheng [1]

1 School of Mechanics and Civil Engineering, China University of Mining and Technology (Beijing), Beijing 100083, China; yanggl531@163.com (G.Y.); bijingjiu@126.com (J.B.); csj24250809@163.com (S.C.)
2 State Key Laboratory for Geomechanics and Deep Underground Engineering (Beijing), Beijing 100083, China
* Correspondence: artisan_lxg@126.com or ZQT1700602028G@student.cumtb.edu.cn

Received: 4 March 2019; Accepted: 26 April 2019; Published: 30 April 2019

Abstract: The impact loading of a notched semi-circular bend (NSCB) specimen of outcrop shale in Changning Sichuan was carried out using a split Hopkinson pressure bar (SHPB) to study the effect of shale bedding on the dynamic crack initiation toughness. Three loading configurations were tested: Crack-divider, Crack-splitter and Crack-arrester loading. Bedding plane has a significant effect on the crack initiation of shale. Under the Crack-divider and Crack-splitter modes, shale had lower dynamic crack initiation toughness. The dynamic crack initiation toughness of the shale was affected by the loading rate for all three loading configurations. The correlation between loading rate and dynamic crack initiation toughness was most significant for the Crack-arrester mode, while the Crack-splitter mode was the weakest. When loading was carried out on Crack-arrester, the bedding plane could change the direction of crack growth. In the Crack-splitter mode, only a small impact energy was needed to achieve effective expansion of a crack. The research results provide a theoretical basis for shale cracking.

Keywords: shale; dynamic crack initiation toughness; NSCB specimen; fracture mode

1. Introduction

China's shale gas resource reserves are huge. Although China has begun commercial development of shale gas, it is still in its infancy and there are many technical difficulties. As a sedimentary rock, shale exhibits bedding characteristics in its structure, which leads to anisotropy in the vertical bedding direction and isotropy in the parallel bedding direction, so shale can be regarded as a transversely isotropic material [1]. The effective cleaving of shale is key to the successful exploitation of shale gas.

Many scholars have studied the physico-mechanical properties of shale and rock with bedding structures. Zhao [2] reported anisotropy of the strength of argillaceous siltstone along the angle between the bedding plane and the principal stress axis. The strength was lowest when the principal stress axis and the weak surface were at an angle of 30°. Mao et al. [3] analyzed the influence of the orientation of slate bedding on development of its compressive strength and failure mechanism. Gao et al. [4] compared the mechanical properties of the slate bedding plane and axial angle in both horizontal and perpendicular directions, and determined the effects of different angles of dip micro-beddings on rock deformation and strength characteristics and parameters. Vernik and Nur [5] studied the anisotropic characteristics of black organic shale wave velocity and concluded that the anisotropy was mainly caused by its microstructure. Kuila et al. [6] studied the anisotropic characteristics of shale caused by a complex stress environment and considered that shale still had high anisotropy characteristics under a high confining pressure. Niandou et al. [7] subjected Tournemire shale to conventional triaxial tests and studied its mechanical behavior under loading and unloading conditions. The results indicated that the shale had significant anisotropic plastic deformation. The failure morphology was strongly dependent on confining pressure and loading bedding angles. Li et al. [8] conducted triaxial compression comparison tests on gas-bearing shales from Barnett, Haynesville, and Eagle Ford in North America and the Lower Silurian Longmaxi Formation in southern China, and determined the mechanical behavior and destruction modes of gas-bearing shales under different stress conditions. Hou et al. [9] used a high-speed camera and acoustic emission system to observe Brazil splitting of shale at different bedding angles. The anisotropic characteristics of shale tensile strength, splitting modulus, and stress peaks were obvious. The extended path of the crack failure surface was strongly influenced by the bedding direction; the acoustic emission activity and energy release were enhanced with an increase of the bedding angle. These results further verified the variation of fracture of shale with bedding angle and the anisotropy of the failure mechanism. Heng et al. [10] performed uniaxial and triaxial compression tests on Longmaxi Formation shale, analyzed the anisotropy of the mechanical properties, strength characteristics, and fracture modes, and revealed anisotropy of the failure mechanism.

In research on the dynamic fracturing of rock, Zhou et al. [11] used a SHPB impact on cracked chevron-notched Brazilian disc (CCNBD) samples and determined the dynamic crack initiation toughness of basalt at different depths of occurrence. Ni et al. [12] carried out dynamic fracture experiments on sandstone single-cleavage drilled compression (SCDC) samples, and determined the dynamic crack initiation toughness of mode I sandstone by experimental–numerical and quasi-static methods. The calculated toughness values were compared. Chen et al. [13] carried out dynamic fracture experiments on Lauren granite NSCB samples. After reaching a dynamic force balance at both ends of the sample, dynamic fracture parameters of mode I were determined, including the dynamic crack initiation toughness, fracture energy, dynamic crack growth toughness and crack growth rate. Dai et al. [14] carried out the dynamic fracture tests of Barre granite [15] NSCB specimens with anisotropy. It was found that the anisotropy of dynamic crack initiation toughness has a loading rate effect, which shows that the anisotropy decreases with the increase of the loading rate.

Many scholars have studied rock materials with bedding structures and dynamic fracture of rock materials under impact loads, but there are few studies on the dynamic fracture of shale under impact loading. In this work, the influence of the bedding direction on mode I dynamic crack initiation toughness and fracture modes of shale was studied using the NSCB specimen recommended by the International Society for Rock Mechanics (ISRM) [16]. The effect of loading rate on dynamic crack initiation toughness was analyzed to provide a theoretical basis for the dynamic cracking of shale.

2. Methods and Materials

2.1. Sample Preparation

Test samples with an obviously developed bedding plane were drilled from outcropping shale of the Changning in Sichuan, which belongs to the shale gas area in Yibin. To study the bedding effect of dynamic fractures of shale in this area, the collected shale blocks were drilled in parallel and perpendicular to bedding directions. According to the ISRM recommendations for rock materials, the NSCB fracture toughness test standard was employed [16]. The samples were first made into Brazilian discs with a diameter of 100 mm and height of 45 mm, then cut and slit to make NSCB samples, as shown in Figure 1. The specific geometric dimensions are shown in Table 1.

Table 1. Geometric parameters of notched semi-circular bend (NSCB) samples.

Diameter D/mm	Thickness B/mm	Pre-Crack Length a/mm	Pre-Crack Width /mm	Pre-Crack Tip Width/mm
100	45	10	0.3	0.1

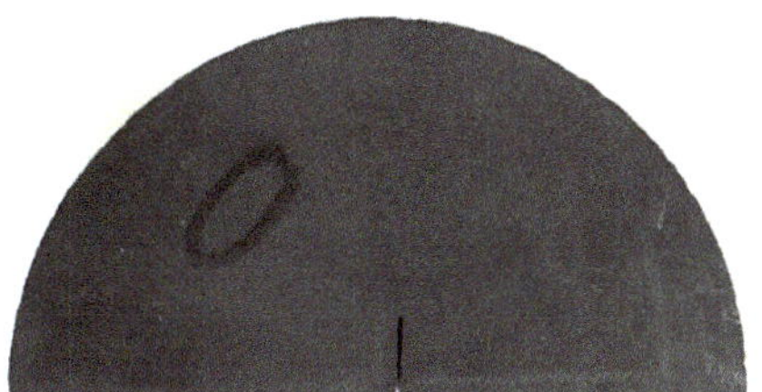

Figure 1. Notched semi-circular bend (NSCB) sample.

2.2. Test System and Principle

The experiments were carried out at the State Key Laboratory of Geomechanics and Underground Engineering, China University of Mining and Technology (Beijing). Dynamic impact loading tests of the NSCB samples were carried out using a 75 mm diameter SHPB experimental system, the parameters of which are shown in Table 2.

Table 2. Split Hopkinson pressure bar (SHPB) experimental system parameters.

Bar Diameter/mm	Incident Bar Length/mm	Transmission Bar Length/mm	Density/(kg/m^3)	Elastic Modulus/GPa	Poisson's Ratio	Longitudinal Wave Velocity /(m/s)
75	3500	3500	7800	210	0.3	5190

According to the one-dimensional stress wave hypothesis of the SHPB experimental technique, a load F_1 is superimposed on the left end of the sample by an incident wave ε_i and reflected wave ε_r. The load F_2 on the right end of the sample is calculated by the transmitted wave ε_t, as shown in Figure 2. These loads are given by:

$$F_1 = AE(\varepsilon_i + \varepsilon_r);\ F_2 = AE\varepsilon_t \tag{1}$$

where A is the cross-sectional area of the bar, E is the elastic modulus of the bar. In the impact loading process, the force balance between the ends of the test sample is the premise for calculating the dynamic crack initiation toughness [13,14,16], i.e., $F_1 = F_2$. The SHPB uses a conical striker to strike the incident bar, which produces ramped (half sine) incident wave, allowing the material to establish a stress balance before it breaks. The entire process of shale fracture was simultaneously recorded during the test by high-speed camera to analyze the anisotropy of the fracture mode.

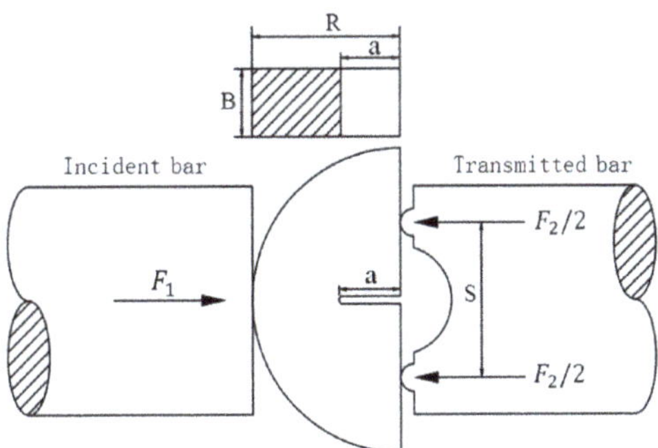

Figure 2. Loading schematic of split Hopkinson pressure bar (SHPB) setup.

According to the International Society for Rock Mechanics (ISRM) test standard [16], the distance between the two supports of the transmitted bar end was set to S = 55 mm, as shown in Figure 2. The pre-crack dimensionless length was defined as $\alpha_a = a/R = 0.2$ and the dimensionless support spacing was $\alpha_s = S/D = 0.55$. By applying a balancing force to the test sample, the pre-crack in the center of the test sample was subjected to mode I fracture. The stress intensity factor of the crack tip in the test sample is given by:

$$K_I(t) = \frac{F(t)S}{BR^{3/2}}Y(\alpha_a) \tag{2}$$

where $F(t)$ is the dynamic load at both ends of the test sample; $Y(\alpha_a)$ is a dimensionless function that depends on the geometric parameters of the prefabricated crack. When $0.15 < \alpha_a < 0.5$, $\alpha_s = 0.55$ and the $Y(\alpha_a)$ function can be expressed as:

$$Y(\alpha_a) = 0.4670 + 3.9094\alpha_a - 8.7634\alpha_a^2 + 16.845\alpha_a^3 \ (\alpha_s = 0.55) \tag{3}$$

According to the basic principles of fracture mechanics [17,18], the instability point of dynamic crack initiation toughness is considered to occur at the maximum load (F_{max}). The engineering significance of the stress intensity factor is used to determine whether brittle fracturing will occur. According to its definition, the critical value of the stress intensity factor is the fracture toughness of the material. Therefore, the dynamic crack initiation toughness of the test sample is obtained by bringing the maximum load on the end face of the test sample into the mode I crack tip stress intensity factor formula recommended by ISRM:

$$K_{Id} = \frac{F_{max}S}{BR^{3/2}}Y(\alpha_a) \tag{4}$$

2.3. Loading Method

Figure 3 provides a schematic diagram of the three loading modes of the NSCB test sample. According to the relative orientations of the loading direction and the bedding structural plane, three loading modes were tested: (a) Crack-divider (b) Crack-splitter and (c) Crack-arrester. Among them, the (a) and (b) loading direction is parallel to the bedding, and the (c) loading direction is perpendicular to the bedding. Each of the loading modes was tested at four different impact pressures to evaluate the dynamic crack initiation toughness and regular rupture of shale under different pressure gradients. Three effective samples which a achieved dynamic force balance were tested for each pressure gradient, as shown in Figure 4.

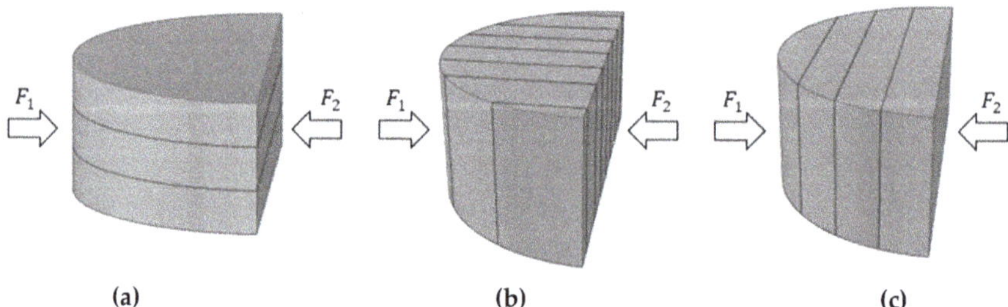

Figure 3. Three loading configurations tested. (**a**) Crack-divider, (**b**) Crack-splitter, (**c**) Crack-arrester.

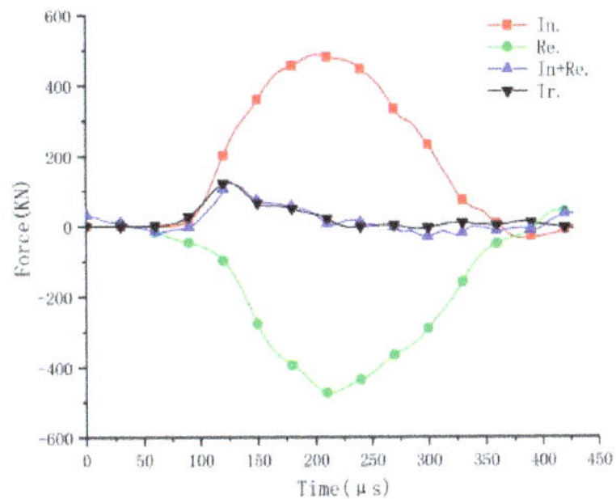

Figure 4. Dynamic force balance (In: Incident wave, Re: Reflected wave, Tr: Transmitted wave).

3. Results and Analysis

3.1. Effect of Bedding on Shale Crack Initiation Toughness

Once the ends of the sample reached a balance of force, the dynamic crack initiation toughness was obtained according to the calculation formula. The loading rate depends on the impact velocity of the SHPB bullet. The calculation results are shown in Tables 3–5 for the three loading configurations.

Table 3. Dynamic crack initiation toughness of shale loading with Crack-divider.

Impact Pressure (MPa)	Test Sample Number	Loading Rate (10^4 MPa $\sqrt{m}$/s)	Dynamic Crack Initiation Toughness K_{Id} (MPa $\sqrt{m}$)	Average Dynamic Crack Initiation Toughness (MPa $\sqrt{m}$)
0.54	P_1-1	21.65	11.53	12.17
	P_1-2	22.36	12.38	
	P_1-3	23.72	12.61	
0.56	P_1-4	25.33	13.65	13.66
	P_1-5	25.26	13.60	
	P_1-6	25.64	13.72	
0.58	P_1-7	28.15	14.38	14.90
	P_1-8	28.72	14.87	
	P_1-9	29.39	15.44	
0.60	P_1-10	33.75	16.67	16.28
	P_1-11	32.51	16.14	
	P_1-12	32.48	16.03	

Table 4. Dynamic crack initiation toughness of shale loading with Crack-splitter.

Impact Pressure (MPa)	Test Sample Number	Loading Rate (10^4 MPa $\sqrt{m}$/s)	Dynamic Crack Initiation Toughness K_{Id} (MPa $\sqrt{m}$)	Average Dynamic Crack Initiation Toughness (MPa $\sqrt{m}$)
0.54	P_2-1	20.50	10.13	9.87
	P_2-2	20.11	9.87	
	P_2-3	19.51	9.60	
0.56	P_2-4	23.41	11.23	11.31
	P_2-5	23.06	11.14	
	P_2-6	24.56	11.56	
0.58	P_2-7	27.20	12.00	11.99
	P_2-8	27.40	12.08	
	P_2-9	26.71	11.89	
0.60	P_2-10	30.60	12.57	12.80
	P_2-11	30.12	12.24	
	P_2-12	31.07	13.58	

Table 5. Dynamic crack initiation toughness of shale loading with Crack-arrester.

Impact Pressure (MPa)	Test Sample Number	Loading Rate (10^4 MPa $\sqrt{m}$/s)	Dynamic Crack Initiation Toughness K_{Id} (MPa $\sqrt{m}$)	Average Dynamic Crack Initiation Toughness (MPa $\sqrt{m}$)
0.54	P_3-1	21.64	12.67	13.32
	P_3-2	22.41	13.43	
	P_3-3	23.24	13.87	
0.56	P_3-4	25.82	15.61	15.75
	P_3-5	26.63	16.40	
	P_3-6	25.57	15.24	
0.58	P_3-7	33.37	19.71	18.97
	P_3-8	32.62	18.78	
	P_3-9	32.17	18.42	
0.60	P_3-10	38.06	23.77	22.51
	P_3-11	36.55	21.59	
	P_3-12	37.23	22.18	

Figure 5 shows the relationship between dynamic crack initiation toughness and loading rate of shale for the three loading modes.

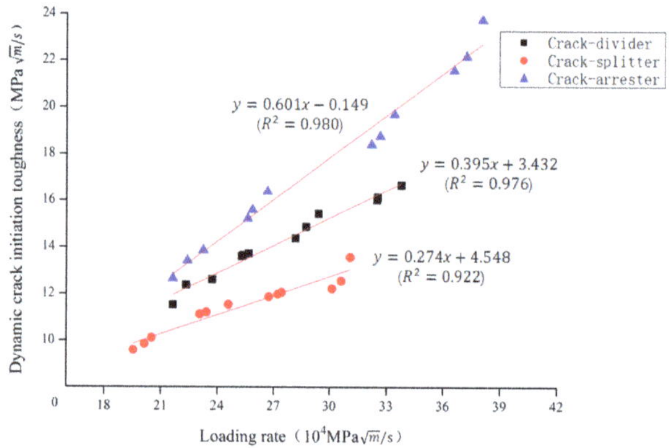

Figure 5. Relationships between dynamic crack initiation toughness of shale and loading rate under different loading configurations.

Under the same loading conditions, the average dynamic crack initiation toughness values of the three sets of shale NSCB specimens increased in the order Crack-splitter < Crack-divider < Crack-arrester. When the loading direction was perpendicular to the bedding plane, the shale cracking mainly depended on the strength of the rock mass and the influence of the bedding plane was almost negligible, so the dynamic crack initiation toughness was at a maximum. When the loading direction was parallel to the bedding plane, the strength of the weak bedding planes had a significant effect on the dynamic crack initiation toughness. Comparing the two parallel loading modes, under Crack-divider loading, the bedding plane had a partial decrease strength effect, the comprehensive strength of the shale in this direction was reduced, and cracking was easier than for the perpendicular bedding plane loading; for Crack-splitter loading conditions, the sample was completely cracked along the bedding plane direction and its strength mainly depended on the strength of the weak bedding planes. Therefore, its dynamic crack initiation toughness was at a minimum. The test results demonstrated that, after clearing the regular bedding plane distribution of shale, the Crack-splitter loading method achieved better cracking with less energy.

Comparing the three fitting curves in Figure 4 showed that the Crack-splitter had the lowest slope, the slope of the Crack-divider was intermediate, and the Crack-arrester had the maximum slope. The different slopes reflect the loading rate correlation of the samples under different loading conditions. Under Crack-splitter conditions, the strength of the test sample mainly depended on the consolidation between bedding planes. There was no obvious correlation of the loading rate for the consolidation between bedding planes. Under Crack-arrester conditions, the dynamic crack initiation toughness increased significantly with the increase of the loading rate, which indicated that the strength of the shale material played a decisive role. Its regular fracture was consistent with correlations for the loading rates of ordinary rock materials.

3.2. Shale Fracture Mode

The NSCB samples showed different fracture modes in the three sets of experiments due to the shale bedding plane. Figure 6 displays three typical fracture modes of the specimens, as recorded by a high-speed camera.

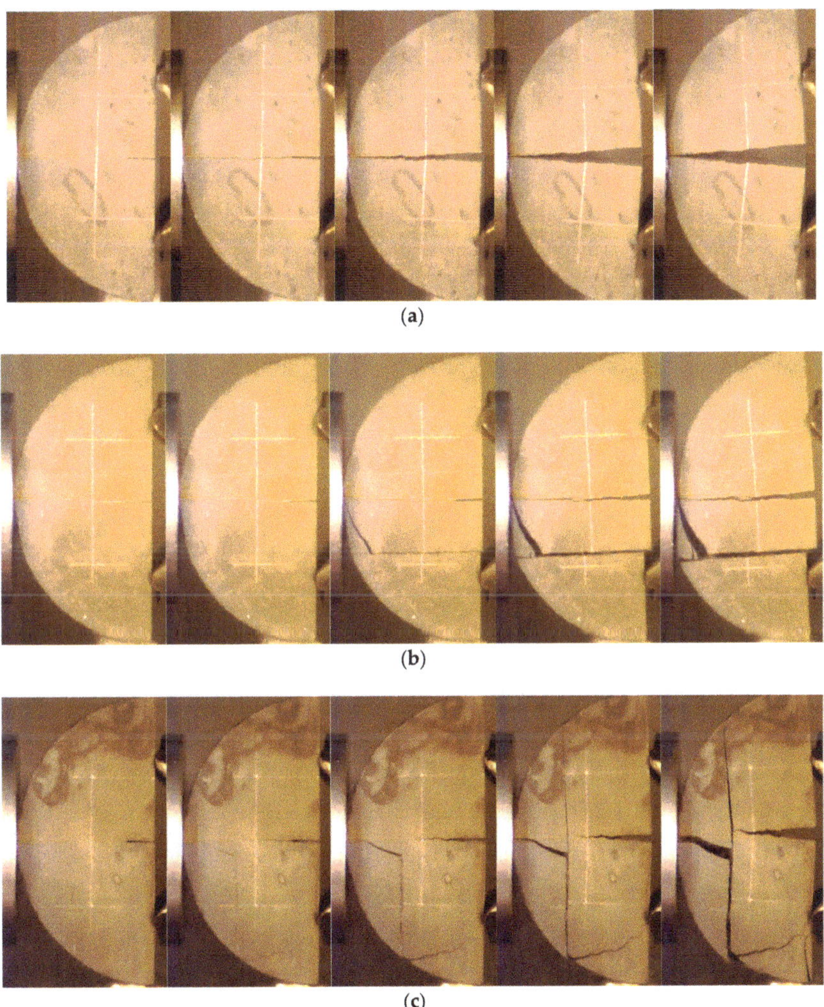

Figure 6. Fracture processes of shale samples for different loading configurations. (**a**) Crack-divider; (**b**) Crack-splitter; (**c**) Crack-arrester.

In the Crack-divider loading mode, the bedding planes had little influence on the crack initiation and the crack expansion path extended along the pre-crack. The crack was straight and no secondary crack appeared. In the Crack-splitter loading mode, in addition to the pre-crack initiation, a new crack appeared along the weak bedding surface. The two crack expansion paths were essentially parallel. In the Crack-arrester mode, the bedding plane was perpendicular to the loading direction: the pre-crack first cracked in the loading direction, followed by cracking of the bedding plane in the vertical loading direction with a faster expansion, and a branching crack formed at the loading point.

By analyzing the above shale fracture results, it was found that Crack-divider loading has smaller dynamic crack initiation toughness than Crack-arrester loading; cracking was easier, but only a single crack formed and no effective crack network, which made the shale gas spread more easily, was generated. This is not conducive to the formation of shale gas diffusion channels. An effective crack network was formed by Crack-splitter and Crack-arrester loading methods. The dynamic crack initiation toughness of the Crack-splitter loaded shale was the minimum of the three configurations and was the easiest configuration to crack.

Shale is a transverse homogenous material. The difficulty of fracture, that is, the size of dynamic crack initiation toughness and fracture modes are key to the success of shale gas development. These research results provide theoretical support for shale gas development.

4. Discussion

The granite used in Dai et al., exhibited strong anisotropy due to its pre-existing micro-cracks induced by long-term tectonic loading. The index of the mode-I dynamic crack initiation toughness anisotropy of the granite is the ratio of the maximum toughness to the minimum toughness of different specimens under the same loading condition, the smaller this ratio, the smaller the anisotropy. We think the bedding structure of the shale used in this experiment is different from the micro-cracks of granite. The two kinds of rocks are different materials with different structural features. Due to the large distribution of graptolite in bedding plane of shale, the consolidation between bedding planes is weaker, and the strength of the rock mass is larger, so it exhibits different anisotropy from the granite.

5. Conclusions

1. We studied the influence of bedding plane on dynamic crack initiation toughness of shale. Under the condition of parallel bedding plane loading, the bedding plane had a significant influence on the shale cracking. The shale had the lowest crack initiation toughness under Crack-splitter loading. Compared with the bedding plane and perpendicular loading directions, the shale cracking mainly depended on the strength of the rock mass, so the dynamic crack initiation toughness had a high value.
2. We analyzed the effect of the loading rate on shale dynamic crack initiation toughness under three loading modes. The Crack-arrester loading mode had the strongest loading rate correlation and Crack-splitter had the weakest.
3. We analyzed the fracture modes of shale under different loading conditions. When loading was carried out using Crack-arrester, the bedding planes can change the crack expansion direction and consume the most energy. The Crack-splitter loading only required a small amount of energy to achieve effective crack expansion.

Author Contributions: G.Y. and X.L. conceived and designed the experiments; S.C. performed the experiments; X.L. and J.B. analyzed the data; X.L. wrote the paper.

Funding: This paper was supported by the opening project of State Key Laboratory of Explosion Science and Technology (Beijing Institute of Techmology). The opening project number was KFJJ19-10M.

Conflicts of Interest: The authors declare no conflict of interest.

References

1. Lee, Y.K.; Pietruszczak, S. Application of critical plane approach to the prediction of strength anisotropy in transversely isotropic rock masses. *Int. J. Rock Mech. Min. Sci.* **2008**, *45*, 513–523. [CrossRef]
2. Zhao, W.R. Strength properties of anisotropic rock of an argillaceous siltstone. *Chin. J. Geotech. Eng.* **1984**, *1*, 32–37.
3. Mao, H.J.; Yang, C.H. Study on influence of discontinuities on mechanical characters of slate. *Chin. J. Rock Mech. Eng.* **2005**, *20*, 53–58.
4. Gao, C.Y.; Xu, J.; Li, Z.H.; Deng, J.H. Experimental study of anisotropically mechanical characteristics of sandy slate in Xuefeng Mountain Tunnel. *Rock Soil Mech.* **2011**, *32*, 1360–1364.
5. Vernik, L.; Nur, A. Ultrasonic velocity and anisotropy of hydrocarbon source rocks. *Geophysics* **1992**, *57*, 727–735. [CrossRef]
6. Kuila, U.; Dewhurst, D.N.; Siggins, A.F.; Raven, M.D. Stress anisotropy and velocity anisotropy in low porosity shale. *Tectonophysics* **2011**, *503*, 34–44. [CrossRef]
7. Niandou, H.; Shao, J.F.; Henry, J.P.; Fourmaintraux, D. Laboratory investigation of the mechanical behaviour of Tournemire shale. *Int. J. Rock Mech. Min. Sci.* **1997**, *34*, 3–16. [CrossRef]
8. Li, Q.H.; Chen, M.; Jin, Y. Experimental research on failure modes and mechanical behaviors of gas-bearing shale. *Chin. J. Rock Mech. Eng.* **2012**, *31*, 3763–3771.
9. Hou, P.; Gao, F.; Yang, Y.G.; Zhang, Z.Z.; Gao, Y.N.; Zhang, X.X.; Zhang, J. Effect of bedding plane direction on acoustic emission characteristics of shale in Brazilian tests. *Rock Soil Mech.* **2016**, *37*, 1603–1612.
10. Heng, S.; Yang, C.H.; Guo, Y.T.; Wang, C.Y.; Wang, L. Influence of bedding planes on hydraulic fracture propagation in shale formations. *Chin. J. Rock Mech. Eng.* **2015**, *34*, 228–237.
11. Man, K.; Zhou, H.W. Research on dynamic fracture toughness and tensile strength of rock at different depths. *Chin. J. Rock Mech. Eng.* **2010**, *29*, 1657–1663.
12. Ni, M.; Gou, X.P.; Wang, Q.Z. Test method for rock dynamic fracture toughness using single cleavage drilled compression specimen impacted by split Hopkinson pressure bar. *Eng. Mech.* **2013**, *30*, 365–372.
13. Chen, R.; Xia, K.; Dai, F.; Lu, F.; Luo, S.N. Determination of dynamic fracture parameters using a semi-circular bend technique in split Hopkinson pressure bar testing. *Eng. Fract. Mech.* **2009**, *76*, 1268–1276. [CrossRef]
14. Dai, F.; Xia, K. Laboratory measurements of the rate dependence of the fracture toughness anisotropy of Barre granite. *Int. J. Rock Mech. Min. Sci.* **2013**, *60*, 57–65. [CrossRef]
15. Goldsmith, W.; Sackman, J.L.; Ewert, C. Static and dynamic fracture strength of Barre granite. *Int. J. Rock Mech. Min. Sci.* **1976**, *13*, 303–309. [CrossRef]
16. Zhou, Y.X.; Xia, K.W.; Li, X.B.; Li, H.B.; Ma, G.W.; Zhao, J.; Zhou, Z.L.; Dai, F. Suggested method for determining the dynamic strength parameters and mode-I fracture toughness of rock materials. *Int. J. Rock Mech. Min. Sci.* **2012**, *49*, 105–112. [CrossRef]
17. Tang, C.A. *Catastrophe in Rock Unstable Failure*; China Coal Industry Publishing House: Beijing, China, 1991.
18. Fan, T.Y. *Dynamic Fracture Mechanics Principle and Its Application*; Beijing Institute of Technology Press: Beijing, China, 2006.

© 2019 by the authors. Licensee MDPI, Basel, Switzerland. This article is an open access article distributed under the terms and conditions of the Creative Commons Attribution (CC BY) license (http://creativecommons.org/licenses/by/4.0/).

Article

Shales Leaching Modelling for Prediction of Flowback Fluid Composition

Andrzej Rogala *, Karolina Kucharska and Jan Hupka

Department of Process Engineering and Chemical Technology, Faculty of Chemistry, Gdansk University of Technology, Narutowicza 11/12, 80-233 Gdansk, Poland; karolina.kucharska@pg.edu.pl (K.K.); jhupka@pg.edu.pl (J.H.)

* Correspondence: andrzej.rogala@pg.edu.pl; Tel.: +48-58-347-2866

Received: 11 February 2019; Accepted: 8 April 2019; Published: 11 April 2019

Abstract: The object of the paper is the prediction of flowback fluid composition at a laboratory scale, for which a new approach is described. The authors define leaching as a flowback fluid generation related to the shale processing. In the first step shale rock was characterized using X-ray fluorescence spectroscopy, X-ray diffractometry and laboratory analysis. It was proven that shale rock samples taken from the selected sections of horizontal well are heterogeneous. Therefore, the need to carry a wide range of investigations for highly diversified samples occurred. A series of leaching tests have been conducted. The extracts were analyzed after leaching to determine Total Organic Carbon and selected elements. For the results analysis significant parameters were chosen, and regression equations describing the influence of rocks and fracturing fluid parameters on the flowback fluid composition were proposed. Obtained models are described by high values of determination coefficients with confidence coefficients above 0.99 and a relatively low standard deviation. It was proven that the proposed approach regarding shale leaching can be properly described using shale models at a laboratory scale, however scaling up requires further investigations.

Keywords: fracturing fluid; flowback fluid; leaching; solid-liquid extraction

1. Introduction

Shales are fine-grained, fissile and the most common sedimentary rocks found in the Earth's crust, composed of clay and other minerals, especially quartz and calcite, as well as organic matter [1]. Shales with proper mature organic matter have gas potential and represent unconventional gas reservoirs. The division into conventional and unconventional reservoirs is associated primarily with the permeability of the reservoir. Conventional reservoirs have a permeability above 0.1 mD, while unconventional reservoirs are below 0.1 mD, and for gas-bearing shales, even below 0.001 mD [2]. The differences in the construction of the reservoir (containing gas in the interstitial space) of a conventional and unconventional shale type are presented in Figure 1. The yellow area represents the gas field space whereas grey/white areas represent rock.

High heterogeneity regarding mineralogical structure, elemental composition and reservoir parameters are typical for shale formations. Individual properties may differ substantially even within individual formations [3–5]. Clear differences in the properties of the deposit occur on a scale of several hundreds to even a few meters [6,7]. The high heterogeneity of the deposit makes all of the well operations diverse and dependent on the given formation. Therefore, a number of deposit stimulation technologies have been developed [8,9].

Figure 1. Schematic structure of conventional (**left**) and unconventional (**right**) gas reservoir.

In the shale rock gas-filled spaces are not interconnected, and the permeability is very low. To ensure gas flow, it is necessary to create a grid of fractures. For this purpose, stimulation of the reservoir is needed, wherein the hydraulic fracturing is the main technology currently applied [10,11]. Hydraulic fracturing consists of using a fracturing fluid pumped under pressure in the range of 700–1200 bar for gas or/and oil-bearing formations. The amount of water needed for fracturing is usually between 10–25 thousand cubic meters and depends on the length of the horizontal part of the borehole. The fracturing pressure must be higher than the tensile strength of the rock, but it cannot exceed the strength parameters of the piping, as it is limited by the power of the pumps. The purpose of the fracturing fluid is primarily to create as much contact surface of the deposit with the well by creating a network of gaps, and to prevent the fractures from closing after fracturing [12]. Fracturing fluids usually consist of at least 90% water containing a proppant as well as other additives [13]. Proppant is a small grain material added to the fracturing fluid in order to prevent the fractures from closing after pressure reduction resulting from hydraulic crushing of rocks [14]. Other additives, which are typically introduced at a level of about 0.5% (v/v), are used to modify the properties of the fluid to enable better penetration of the formation and provide compatibility between the fluid and reservoir [12,15].

After the fracturing is completed, 10–50% of the fluid returns automatically or by stimulation to the surface. The returning stream is called flowback fluid. In extreme cases, the flowback fluid does not return to the surface at all, or returns in larger quantities. The rest of the fluid remains in the reservoir [16]. Flowback fluid differs significantly from the fracturing fluid, as it contains suspended fine rocks as a result of the leaching process, reservoir water, drilling muds, and also various chemical substances, such as dissolved solids (e.g., chlorides, sulfates, etc., measured as total dissolved solids or TDS), suspended solid particles (TSS), bacteria, metals (e.g., calcium, magnesium, barium, strontium), iron compounds, aromatic hydrocarbons, carbon dioxide, hydrogen sulphide and other components [17].

Average results of fracturing and post-treatment fluid analysis from the Marcellus Shale basin in the United States are presented in Table 1. The results of the concentrations of the ingredients may vary by a factor of several dozen with the concentrations of the fracturing fluid [18].

Although the fracturing fluid is environmentally friendly, flowback fluid should be treated as an environmental threat due to its composition. It can have adverse effects on the environment, and hence requires purifying and/or disposal. Flowback fluid, treated as a mining waste, can be transported to another drilling plant, and also be transferred to companies dealing with the disposal of mining waste. However usually it is transferred to a sewage treatment plant or another installation that neutralizes waste [19,20]. Due to the high variability of the flowback fluid and the difficulties of its composition estimation, models and tools that help in its prediction are desirable.

Table 1. Comparison of concentrations of the most important components in fracturing fluid and flowback fluid from the Marcellus Shale [18].

Component	Fracturing Fluid		Flowback Fluid	
	Range of Concentrations (ppm)	Median Concentration (ppm)	Range of Concentrations (ppm)	Median Concentration (ppm)
Na^+	25.70–6190	67.8	10700–65100	18000
Ca^{2+}	6.70–2990	32.9	1440–23500	4950
Mg^{2+}	1.20–235	6.7	135–1550	559
Fe^{2+}	0.00–14.3	1.2	10.8–180	39
Ba^{2+}	0.06–87.1	0.4	21.4–13900	686
Cl^-	4.10–3000	42.3	26400–148000	41850
HCO_3^-	<1.00–188	49.9	29.8–162	74.4
NH_4^+	0.58–441	5.9	15–242	82.4

Flowback fluid analysis is being widely used in shale gas recovery modelling. There are many models describing propagation of shale fracture to assess possible production of shale gas [21–26]. Some attempts to describe flowback fluid production mechanisms are reported [27–30] and several methods to develop volume control are described [31,32]. Unfortunately, there are not many investigations dedicated to the prediction of the composition of fracturing flowback fluid and hardly any tools to limit the migration of chosen chemical compounds from the reservoir to the fluid. Methods focused on computer programs, supported by precipitation tests [33], statistics of reservoir data [34] or statistics of fracturing wells in chosen basin of gas bearing shales can be specified. However evaluation is not possible because the proposed tools are often unavailable and not described enough to test them due to the unnormalized character of the data [35]. Nevertheless very useful information can be found on the basis of Total Organic Carbon (TOC) regarding salt migration into flowback fluid streams [33]. Also ionic strength and the influence of ion interactions exhibit a crucial role [35]. Due to the unnormalized character of the data presented in the literature, the models are often non-comparable. The most interesting modeling approaches are presented below.

A promising model was presented by Gdanski et al. [35]. The model allows one to predict the amount and composition of the flowback fluid. A two-dimensional numerical modelling was used, taking into account the physics of fracturing fluid flow and flowback fluid and chemical interactions in the borehole. Leaching of sodium, potassium, chlorides, sulphates, carbonates and boron were considered in the model. The model provides separate data for post-treatment fluids and reservoir waters. Based on the adjustment of the fluid production rate, it is also possible to estimate certain properties of the reservoir, such as relative permeability and capillary pressure. However, it should be noted that the author based his model on a commercial simulator [36], that is unfortunately commonly unavailable and therefore, impossible to evaluate. In addition, the conducted considerations concerned on small-scale fracturing in a vertical hole with accurate characteristics. Such a large amount of data cannot be obtained for full-scale horizontal wells with diversified fluids, therefore, the model is unapplicable for the conditions presented by other types of wells.

Another model was developed by the Barbot team [33,34]. It concerned the fracturing of wells in the Devonian shale in the north-eastern United States. An analysis of 160 flowback fluids was used to build a model for the prediction of the quantity and composition of the flowback fluids in the American Marcellus basin. Models for the concentration prediction of sodium, calcium, magnesium, barium, strontium, bromides and chlorides were developed. The performed tests may be applicable during the management of the flowback fluid on this deposit. Despite the authors' conclusions, analysis of only 160 flowback fluid samples cannot give enough data for modelling with no proper relation to shale rock properties. This type of model can be built at the advanced stage of shale gas extraction only for very specific areas and it is impossible to extend it for other reservoirs.

A much more advanced model was presented by Balashov [37] In addition to analyzing a larger amount of data, Balashov et. al. included considerations regarding the diffusion model based on shale rock. A numerical model was created and calibrated using field data from a specific area. In the conclusions, the author claimed that well-fitting numerical models were obtained, unfortunately we cannot verify this, as the model itself has not been sufficiently presented.

Unfortunately, the approaches presented in the literature are completely different and there is a lack of available data to evaluate and compare them. Therefore, an idea to create a new type of model for flowback fluid composition prediction was proposed and presented in this paper. The authors will make an attempt to contribute to the understanding of leaching method development.

A new approach to modeling flowback fluid composition prediction has been proposed. Research was carried out to describe the mechanism of flowback fluid production and composition generated during leaching of shale rock by fracturing fluid. The process was considered in a state of equilibrium due to the long residence time of fracturing fluid in the reservoir. Leaching course depends mainly on physical parameters like the heterogeneity of solids, size of particles, porosity, permeability and temperature of the process as well as pH, conductivity, total organic carbon (abr. TOC), time complexed reactions, redox potential and biological activity [38]. For shale leaching applying complexed fracturing fluids, a model complexity limiting approach needs to be applied. It was proposed to find a model based on rock properties which cannot be modified. Statistically significant parameters for leaching were identified as follows: pH, temperature, TOC and ionic strength [39–41], and the mentioned paramaters can be modified by changing the fracturing fluid composition. To make the model applicable, shale rock particles from drilling mud logging were prepared and used. Application of drill cuttings from mud logging can cause problems with sample preparation [42]. Nevertheless, at the same time, cross-sectional material from the fracturing process with no need for much expensive coring may come in handy. The proposed approach will allow other scientific groups and laboratories to evaluate the presented data and extend the presented considerations using samples from other different shale gas reservoirs. An overall conceptualization of data generation and processing is presented in Figure 2.

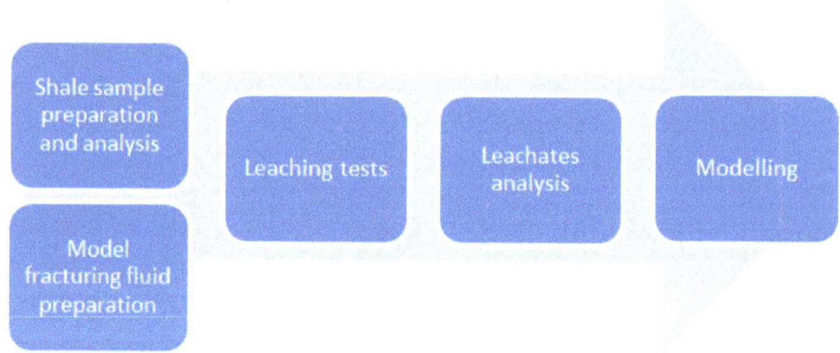

Figure 2. Overall conceptualization of data generation and processing.

2. Materials and Methods

2.1. Shale Sample Acquisition

Samples of drill cuttings for further research were obtained from bentonite drilling mud. The samples come from the Baltic basin, in the Pomeranian area (Pomeranian voivodeship, Poland), from a depth of about 4.250 meters, taken along the entire length of the horizontal section (120 meters each).

2.2. Shale Sample Preparation

In order to remove the mud from the surface of the cuttings, cleaning of the tested material was carried out. For this purpose, the shale sample was sprinkled with a stream of water, next different sieves were used (sieve sizes: 2.362, 0.18, 0.088 mm). Samples were separated until the mud was completely removed. According to Carugo et al.'s [43] recommendations, a 2.362–0.18 mm size fraction was selected for further investigations. The higher fraction represents cuttings that could move from other parts of the well, or come from a damaged well walls, smaller drill cuttings were too small to perform further research due to their properties. After cleaning and sieving, the samples were dried at 40 °C for 24 h. The dried samples were ground in a vibratory ball mill to achieve a grain size below 0.088 mm, and then reduced by a flattened pile.

2.3. Shale Samples Analysis

Mineralogical characterization was realized using X-ray Diffraction (XRD) using a Miniflex 600 X-ray diffractometer (Rigaku, Neu-Isenburg, Germany). The semi-quantitative mineral composition was calculated using Rietveld method, dedicated to rocks with high concentration of clay minerals. Elemental analysis was performed using an X-Ray fluorescence (XRF) spectrometer (S8 TIGER Wavelength Dispersive X-ray Fluorescence WDXRF, Bruker, Billerica, MA., USA). Total organic carbon was analyzed using a CHNS elemental analyzer (Flash, 2000; Thermo, Waltham, UK).

2.4. Preparation of Model Fracturing Fluids

In order to prepare model fracturing fluids, acetate (pH = 5), dihydrogen phosphate (pH = 7) and ammonium (pH = 9) buffers with appropriate acid-base ratios were used. The prepared fluid was buffered at 0.002 M. It results from the analysis of the components of fracturing fluids used in Polish deposits. Also, ionic strength and total organic carbon range is a result of calculations based on the compositions of these fracturing fluids. The ionic strength was controlled by the addition of potassium chloride and the organic carbon content added as propylene glycol.

2.5. Leaching Tests

Leaching tests were carried out according to OECD 106: Adsorption - Desorption method [42] using 10 g of a shale sample per 100 mL of "model fracturing fluid" (MFF) which is a laboratory composed model fluid to conduct leaching tests. Liquid/solid phase separation was preformed via centrifugation at 4000 rpm, vacuum filtration and final separation with 0.45 μm pore size polypropylene membranes. Leaching tests were conducted according to parameters presented in Table 2 which contains a summarized plan listing the following variables: temperature, pH, ionic strength and TOC.

Table 2. Parameters of "model fracturing fluids" (MFF) for leaching tests of shale samples.

No.	MFF	t (°C)	pH	IS (mol/L)	TOC (g/L)
		X1	X2	X3	X4
1	P1	60	9	0.51	0.06
2	P2	60	9	0.03	4.06
3	P3	60	5	0.51	4.06
4	P4	60	5	0.03	0.06
5	P5	80	9	0.51	4060
6	P6	80	9	0.03	0.06
7	P7	80	5	0.51	0.06
8	P8	80	5	0.03	4.06
9	P9	80	7	0.51	4.06
10	P10	60	7	0.03	0.06
11	P11	80	9	0.51	2.06
12	P12	60	5	0.03	2.06
13	P13	70	7	0.27	2.06
14	P13	70	7	0.27	2.06
15	P13	70	7	0.27	2.06

t—temperature of the experiment equal to assumed wellbore conditions, pH—pH of the MFF, IS—ionic strength of the MFF, TOC—total organic carbon of MFF.

For each of 10 rock samples 15 tests were performed—13 different fluids and two repetitions for fluid P13 (for the statistical approach) which is the average value of the range of the analyzed variables. The range of temperatures chosen for leaching tests is based on field data from service companies in Poland (i.e. BNK Petroleum, Camarillo, CA, USA) and corresponds with real temperatures in the depth of fractured horizontal sections of the well from Polish reservoirs with the highest gas production potential. Unfortunately, the tests presented are destructive to the core samples. Therefore, it is highly recommended to use shale cuttings which are cheaper and more available in comparison with core samples. Leachate compositions was analyzed by ICP-OES using an iCAP 6500 Duo spectrometer (Thermo, Waltham, UK). Data analysis and other calculations were performed using the RStudio Desktop (v. 1.0.143) software. Laboratory data obtained during analyses of liquids after leaching were grouped and entered into the program as descriptive variables. Temperature, ionic strength, pH, TOC and possible interactions between them as well as the results of rock analysis were variables. Then RStudio Desktop tools were used to compare different possible models, allowing the selection of the best one. A multistep linear regression was used to obtain the presented models. Only the models for which the coefficient of determination (abr. R^2) was above 0.85 were taken into consideration, the probability value for all describing variables was below 0.05 (variable values are not random), so the confidence level was always above 0.95 (note: R^2 coefficients are calculated for obtained model equations not for the residual analysis diagrams). In the next step only those equations for which all separate levels of confidence for every separate parameter were above 0.95 were considered. Then the model with the highest determination coefficient was chosen for further analysis. For chosen model the standard deviation was always the lowest taking into account considered models and its value justifies the correctness of application of R^2 as a crucial parameter to choose the most adequate model. The residual analysis showed a sufficient number of describing variables (small inclination angle to the abscissa) and there was no correlation between residuals and described variables (Pearson between −0.5 and 0.5). The Pearson coefficients were calculated for the correlation between elements concentrations in leachates and the residuals of the model. The residuals were calculated as the difference between laboratory and model data values. The model was selected basing on the comparison of the standard model error, the determination coefficient, the level of significance and the number of parameters relevant to the correct description of the model. Other parameters presented in the next section were calculated using statistical and mathematical methods implemented in RStudio software [43].

3. Results and Discussion

Results of elemental and mineralogical analysis are presented in Tables 3 and 4. In Table 3, only elements which provided data for creating models are presented.

Table 3. Results of elemental analysis of shale samples (A1 ÷ A10) using the XRF method.

No. of Sample	Element Concentration in the Shale Sample (CS) (ppm)									Aw (m^2/g)	TOC_S (%)
	B	Ba	Li	Mg	Mo	Rb	Si	Sr	Ca		
A1	4050	821	139	18240	10	3372	247285	401	116287	15.17	2.11
A2	3988	593	149	14820	95	3570	278445	175	35020	16.15	0.84
A3	3325	708	97	19570	82	2542	229900	605	179117	12.77	2.41
A4	3637	967	112	15960	9	3469	241965	179	131634	10.24	2.43
A5	3467	1208	110	17670	97	2578	262200	178	114948	10.67	2.72
A6	3694	774	115	23560	97	3369	262105	262	100013	14.05	1.40
A7	3361	1119	88	26695	82	3280	248045	212	123188	14.61	1.61
A8	3918	1455	122	20805	106	3068	273505	178	68804	9.72	4.40
A9	3425	1395	120	19665	100	2777	277400	154	58916	13.29	1.81
A10	3513	842	119	18335	105	3358	282150	140	59019	12.62	1.77

Aw—internal area of shale sample, TOC_S—Total organic carbon in shale sample, B, Ba, Li, etc.—symbols of the indicated elements, CS_X—element X concentration in shale sample.

Table 4. Results of mineralogical analysis of shale samples (A1 ÷ A10) using the XRD method.

No. of Sample	Group of Minerals (%)				
	G1	G2	G3	G4	G5
A1	21.50	11.95	9.13	15.24	42.19
A2	25.25	3.66	0.08	10.74	60.28
A3	18.10	37.19	0.12	20.83	23.75
A4	28.14	25.49	0.25	18.47	27.64
A5	25.94	27.60	0.12	23.84	22.50
A6	29.33	43.72	1.85	6.16	18.94
A7	19.61	46.83	0.24	8.82	24.50
A8	32.52	24.94	0.12	5.98	36.44
A9	46.16	20.36	0.88	9.31	23.29
A10	25.56	21.96	0.12	9.22	43.15

G1—sum of quartz, plagioclase (albite) and potassium feldspar (microcline, orthoclase), G2—sum of calcite and dolomite content, G3—sum of barite and pyrite content, G4—sum of chlorite, illite, montmorillonite and kaolinite content, G5—sum content of muscovite, biotite, paragonite and glauconite.

A detailed mineralogical analysis is attached in the Appendix A (Table A1). The shale samples were tested for organic carbon content. When analyzing the results in Table 4, it can be assumed that the section of the well from which the A8 sample was taken is the most promising in terms of hydrocarbon production. Also, the TOC content for samples A1, A3, A4 and A5, which is above 2%, theoretically indicates that the hydrocarbon content in the test rock is sufficient for gas operation. However, this fact cannot be unambiguously conducted because no thermal maturity studies of organic matter have been carried out.

The analysis of the results of the mineralogical composition proves that the mineral structure of the rocks forming the gas-bearing slate form shows high heterogeneity in the single well's circle. The distance between successive sections was only 120 meters, and the composition of the rocks from each of them is completely different, even for neighboring sections. During hydraulic fracturing, the operator should pay attention to the need to use different fracturing fluid compositions and parameters to obtain an effective gas flow. The big advantage of such a heterogeneity of the deposit in relation to the proposed model is the wide scope of the model's application. The disadvantage is the poor estimation of impurities in the flowback fluid.

The content of minerals from the smectites group (montmorillonite) for samples A3 and A5 and A7 (Table A1 in the Appendix A) is respectively 10.73%, 0.12% and 0.24%, while for the remainder it is equal to 0%. This result means that the formation is heterogeneous on the scale of a single well. Most sections should be slightly susceptible to water. The content of carbonate minerals (which include calcite and dolomite) for the tested samples is in the range of 3.7–46.8%. Such a large variation indicates that the tested samples differ significantly in reactivity, which will certainly have an impact on the conducted leaching tests. The amount of quartz from the analyzed cuttings ranges from 13.6% to 35%. It can be deduced that in individual sections of the borehole, the formation will exhibit varied mechanical properties (diversification of brittleness). The content of other minerals is also very diverse, so the leaching process may proceed with different leaching efficiency for different rocks using the same model fracturing (leaching) fluids.

Differences in specific surface (which is actually the open surface as a result of grinding samples in vibratory mills) probably occur mainly due to differences in hardness and cleavage of the rock and random factors. The reference to the open area is necessary for the quantitative mathematical description of the model to be reliable. In selected cases, the surface may be negligible when the content of the component available through the open surface exceeds the solubility of this component in the fracturing fluid of a given composition and temperature.

The results of elemental analysis of leaching fluids and derived, as equations, models that describe correlation between concentrations of chosen elements (c_B, c_{Ba}, etc.) and variables. Residual analysis (difference between measured and calculated values) and maximum and minimum values calculated for the part of the model that can be influenced by controllable parameters of the fracturing fluid (c'_{min} (60), c'_{min} (80), c'_{max} (60), c'_{max} (80)) are presented in Tables 5–13. Composition of shales and temperature in the borehole are parameters of a particular well, therefore, we can change: IS, TOC and pH (by changing the fracturing fluid composition) so c'_B, is a part of the whole c equation which can be manipulated and is used to calculate the possible range of changeability of the model by fracturing fluid parameter manipulations. The values of c'_{min} and c'_{max} were calculated for two extreme temperature values. It was calculated that the minimum and maximum values for all models are equal to 60 or 80 °C (minimum and maximum temperature). The differences between these values are presented in Table 14. These results indicate how much the concentration of the element in the flowback fluid can be reduced by proper control of the parameters of the fracturing fluid according to the developed models. Due to the statistical nature of the model, in some cases lowering the concentration could give a negative value. This should be taken into account when using and possibly implementing the model using a computer tool. Model fracturing fluids are characterized in Table 14.

Table 5. Results of leaching samples A1 to A10 rocks using fluids P1 to P13 for Boron (B).

B concentration in samples after leaching of A1 to A10 rocks using fluids P1 to P13

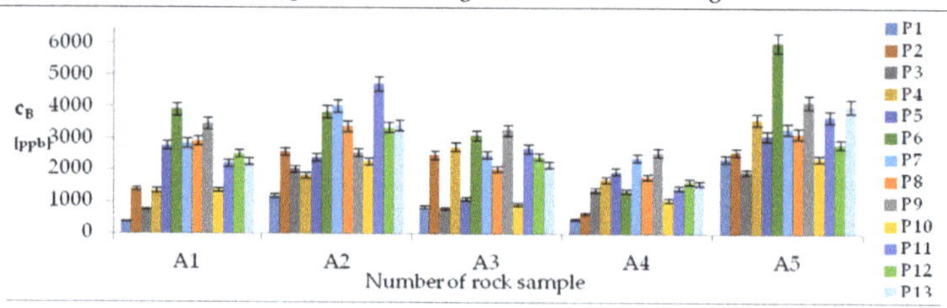

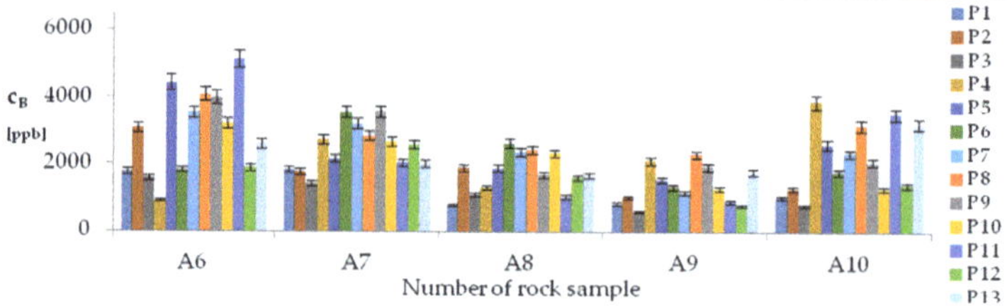

Model

$$c_B = -5155 + 35.6t - 549.7TOC - 74.7(TOC)^2 - 341.7pH \times IS + 25.5pH \times TOC + 21.5t \times IS + 9.3t \times TOC - 211.6G1 - 115.4G2 - 60.2G4 - 183.9G5 + 2.4CS_B - 81CS_{Mo} + 0.06CS_{Si} \pm [286.4]$$

$$c'_B = -35.6t - 549.7TOC - 74.7\,(TOC)^2 - 341.7pH \times IS + 25.5pH \times TOC + 21.5t \times IS + 9.3t \times TOC$$

R^2	0.90	Residual analysis 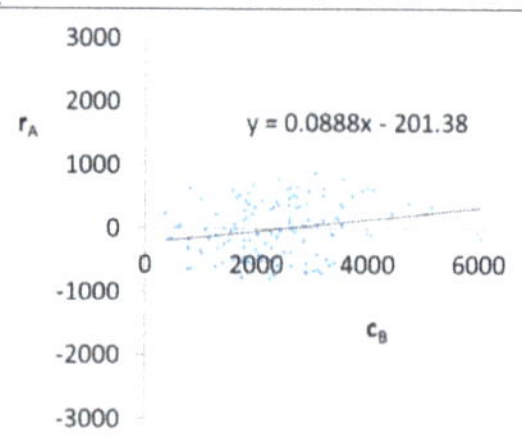
p	<0.000001	
a	0.0888	
α = arctan (a)	5.1°	
Pearson	0.30	
c_{min} (t = 60 °C)	1238 ppb	
c_{max} (t = 60 °C)	1967 ppb	
c_{min} (t = 80 °C)	2181 ppb	
c_{max} (t = 80 °C)	3159 ppb	

c_B—concentration of boron in leachates; r_A-residual in residual analysis; a—directional coefficient; arctan-arcus tangent which is equal to angle between regression line in residual analysis and 0x axis; Pearson-Pearson correlation coefficient for r_A and c_B; c'_{min}(t = 60 °C); c'_{min}(t = 80 °C); c'_{max}(t = 60 °C); c'_{max}(t = 80 °C)—explained before Table 5; CS_B, CS_{Mo}, CS_{Si}- B, Mo, Si concentrations in shale sample.

Table 6. Results of leaching samples A1 to A10 rocks using fluids P1 to P13 for Barium (Ba).

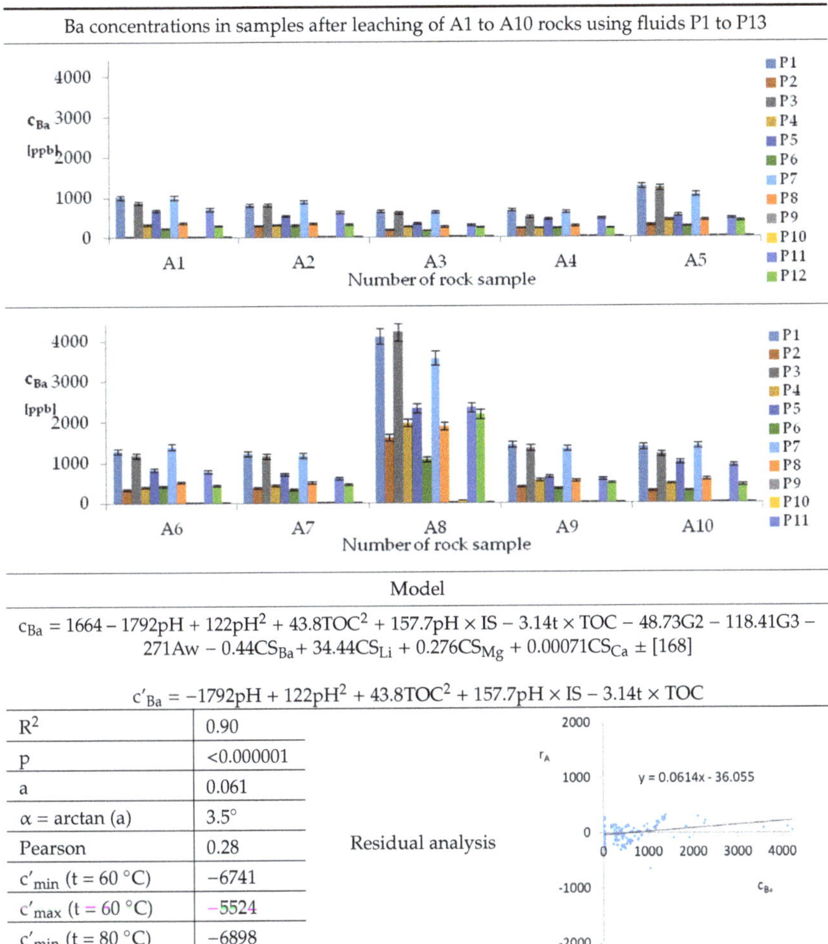

Ba concentrations in samples after leaching of A1 to A10 rocks using fluids P1 to P13

Model

$$c_{Ba} = 1664 - 1792pH + 122pH^2 + 43.8TOC^2 + 157.7pH \times IS - 3.14t \times TOC - 48.73G2 - 118.41G3 - 271Aw - 0.44CS_{Ba} + 34.44CS_{Li} + 0.276CS_{Mg} + 0.00071CS_{Ca} \pm [168]$$

$$c'_{Ba} = -1792pH + 122pH^2 + 43.8TOC^2 + 157.7pH \times IS - 3.14t \times TOC$$

R^2	0.90	Residual analysis
p	<0.000001	
a	0.061	
α = arctan (a)	3.5°	
Pearson	0.28	
c'_{min} (t = 60 °C)	−6741	
c'_{max} (t = 60 °C)	−5524	
c'_{min} (t = 80 °C)	−6898	
c'_{max} (t = 80 °C)	−5527	

c_{Ba}—concentration of barium in leachates; r_A—residual; a—directional coefficient, arctan-arcus tangent; Pearson-Pearson correlation coefficient for r_A and c_B; c'_{min}(t = 60 °C); c'_{min}(t = 80 °C); c'_{max}(t = 60 °C); c'_{max}(t = 80 °C)—explained before Table 5; CS_{Ba}, CS_{Li}, CS_{Mg}, CS_{Ca}-Ba, Li, Mg, Ca concentrations in shale sample.

Table 7. Results of leaching samples A1 to A10 rocks using fluids P1 to P13 for Strontium (Sr).

Sr concentration in samples after leaching of A1 to A10 rocks using fluids P1 to P13

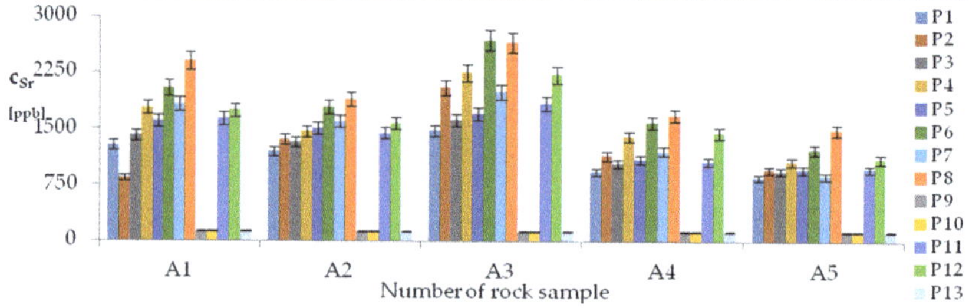

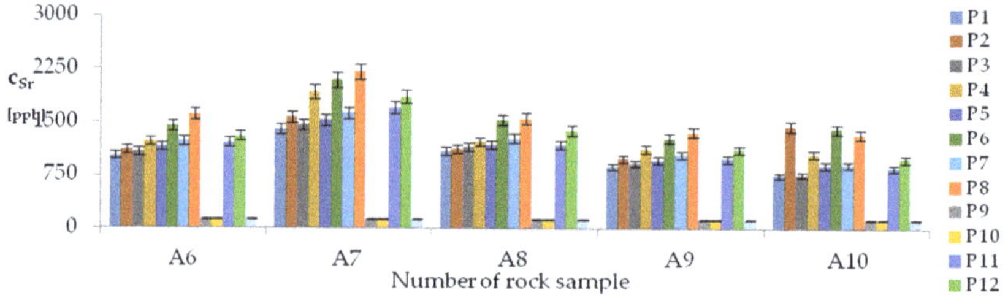

Model

$$c_{Sr} = 11995 - 4001pH + 273.6pH^2 + 12.1TOC^2 + 2.2pH \times t - 4.8pH \times TOC - 7t \times IS - 72.4IS \times TOC + 40.2G2 + 25.6G5 + 129.1Aw + 1.11CS_{Ba} - 0.1CS_{Mg} - 0.005CS_{Si} + 1.5CS_{Sr} \pm [134.6]$$

$$c'_{Sr} = -4001pH + 273.6pH^2 + 12.1TOC^2 + 2.2pH \times t - 4.8pH \times TOC - 7t \times IS - 72.4IS \times TOC$$

R^2	0.95	Residual analysis
p	<0.000001	
a	0.046	
α = arctan (a)	2.6°	
Pearson	0.24	
c'_{min} (t = 60 °C)	−13910 ppb	
c'_{max} (t = 60 °C)	−12700 ppb	
c'_{min} (t = 80 °C)	−13670 ppb	
c'_{max} (t = 80 °C)	−12310 ppb	

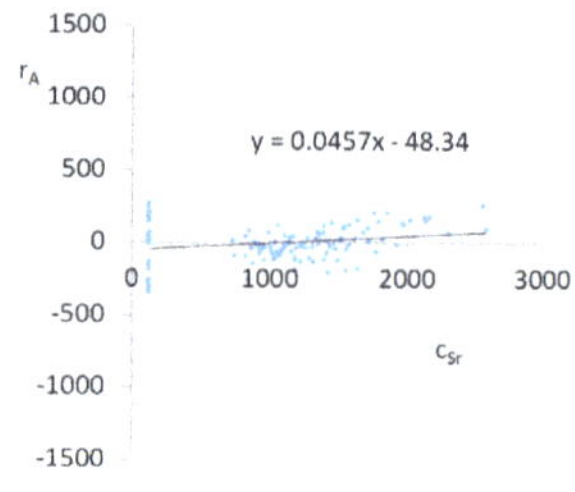

c_{Sr}—concentration of strontium in leachates; r_A—residual; a-directional coefficient; arctan-arcus tangent; Pearson-Pearson correlation coefficient for r_A and c_B; c'_{min}(t = 60 °C); c'_{min}(t = 80 °C); c'_{max} (t = 60 °C); c'_{max}(t = 80 °C)—explained before Table 5; CS_{Ba}, CS_{Li}, CS_{Mg}, CS_{Ca}—Ba, Li, Mg, Ca concentrations in shale sample.

Table 8. Results of leaching samples A1 to A10 rocks using fluids P1 to P13 for Molybdenum (Mo).

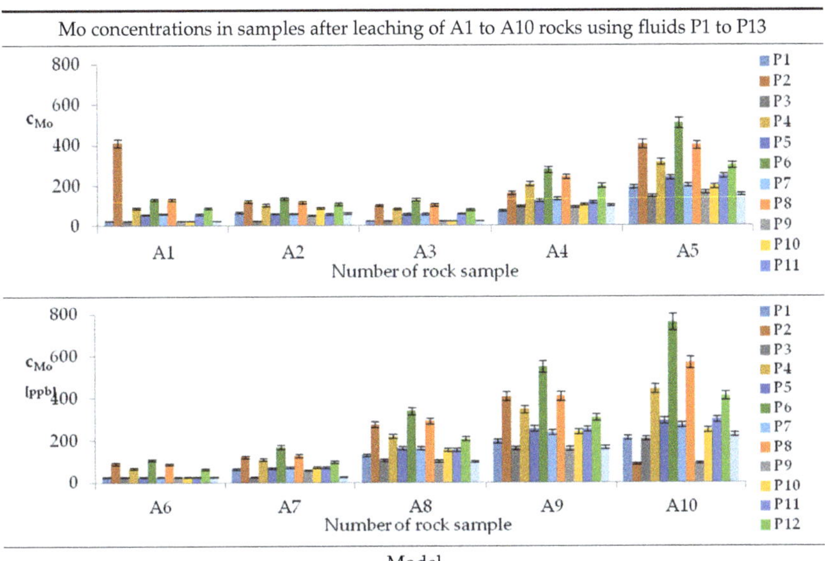

Model

$$c_{Mo} = -1002.7 + 37.82t - 15576IS + 207.9TOC + 17.68pH^2 - 4.05pH \times t + 145.83pH \times IS - 3.59t \times TOC + 166.94IS \times TOC + 3.82G4 + 2.41G5 - 0.38CS_B + 10.19CS_{Mo} - 0.22CS_{Sr} \pm [32.7]$$

$$c'_{Mo} = 37.82t - 15576IS + 207.9TOC + 17.68pH^2 - 4.05pH \times t + 145.83pH \times IS - 3.59t \times TOC + 166.94IS \times TOC$$

R^2	0.91	Residual analysis
p	<0.000001	
a	0.069	
α = arctan (a)	4.0°	
Pearson	0.29	
c'_{min} (t = 60 °C)	1069	
c'_{max} (t = 60 °C)	1695	
c'_{min} (t = 80 °C)	1342	
c'_{max} (t = 80 °C)	1431	

c_{Mo}-concentration of molybdenum in leachates; r_A—residual; a-directional coefficient; arctan-arcus tangent; Pearson-Pearson correlation coefficient for r_A and c_B; c'_{min}(t = 60 °C); c'_{min}(t = 80 °C); c'_{max}(t = 60 °C); c'_{max}(t = 80 °C)—explained before Table 5; CS_B, CS_{Mo}, CS_{Sr}—Ba, Mo, Sr concentrations in shale sample.

Table 9. Results of leaching samples A1 to A10 rocks using fluids P1 to P13 for Rubidium (Rb).

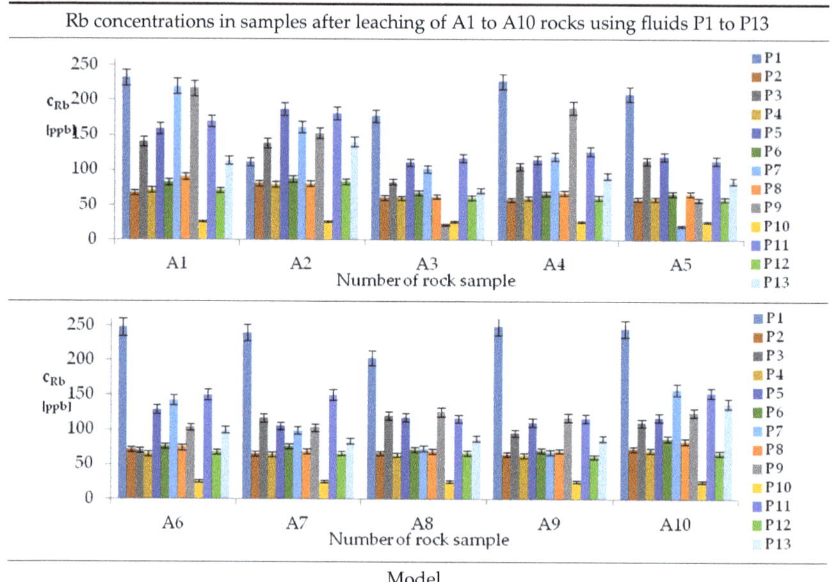

Model

$$c_{Rb} = -77.42 + 1.05t + 462.79\,IS + 27.92\,pH \times IS - 6.49t \times IS - 21.88IS \times TOC - 0.42G2 + 0.024CS_{Rb} \pm [18.71]$$

$$c'_{Rb} = 1.05t + 462.79IS + 27.92pH \times IS - 6.49t \times IS - 21.88IS \times TOC$$

R^2	0.86	Residual analysis
p	<0.000001	
a	0.102	
α = arctan (a)	5.8°	
Pearson	0.31	
c'_{min} (t = 60 °C)	125.9	
c'_{max} (t = 60 °C)	227.5	
c'_{min} (t = 80 °C)	80.6	
c'_{max} (t = 80 °C)	182.2	

c_{Rb}-concentration of rubidium in leachates; r_A—residual; a-directional coefficient; arctan-arcus tangent; Pearson-Pearson correlation coefficient for r_A and c_B; c'_{min}(t = 60 °C); c'_{min}(t = 80 °C); c'_{max}(t = 60 °C); c'_{max}(t = 80 °C)—explained before Table 5; CS_{Rb},–Rb concentrations in shale sample.

Table 10. Results of leaching samples A1 to A10 rocks using fluids P1 to P13 for Lithium (Li).

Li concentrations in samples after leaching of A1 to A10 rocks using fluids P1 to P13

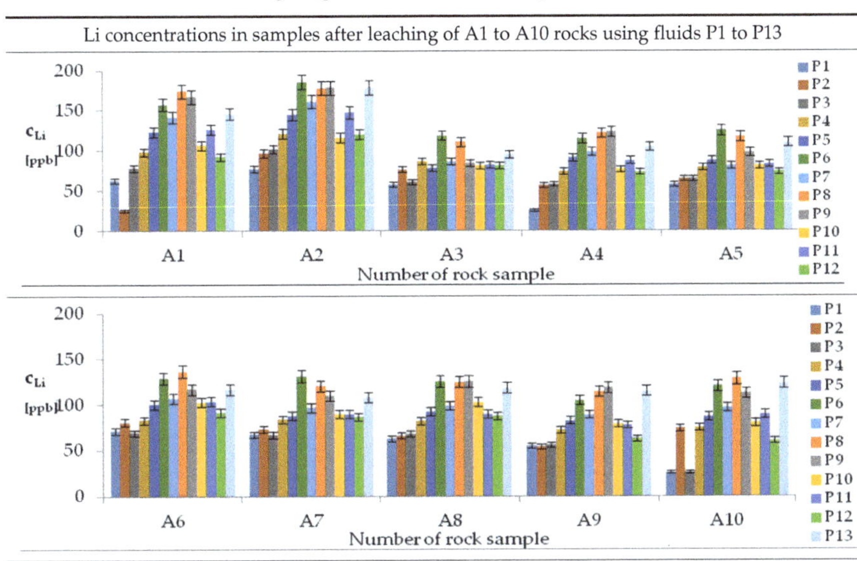

Model

$$c_{Li} = 500.56 + 78.85pH + 150.00IS - 7.51pH^2 + 0.38pH \times t - 2.50t \times IS - 3.53IS \times TOC - 3.03G1 + 1.23G4 - 1.73G5 + 0.41CS_{Li} - 0.0019CS_{Si} + 0.08CS_{Sr} - 0.0018CS_{Ca} \pm [10.66]$$

$$c'_{Li} = 78.85pH + 150.00IS - 7.51pH^2 + 0.38pH \times t - 2.50t \times IS - 3.53IS \times TOC$$

R^2	0.88	Residual analysis	y = 0.0816x - 6.7988 (r_A vs. c_{Li})
p	<0.000001		
a	0.082		
α = arctan (a)	4.7		
Pearson	0.27		
c'_{min} (t = 60 °C)	298.8		
c'_{max} (t = 60 °C)	343.6		
c'_{min} (t = 80 °C)	341.6		
c'_{max} (t = 80 °C)	371.4		

c_{Li}-concentration of lithium in leachates; r_A—residual; a-directional coefficient; arctan-arcus tangent; Pearson-Pearson correlation coefficient for r_A and c_B; c'_{min}(t = 60 °C); c'_{min}(t = 80 °C); c'_{max}(t = 60 °C); c'_{max}(t = 80 °C)—explained before Table 5; CS_{Li}, CS_{Si}, CS_{Sr}, CS_{Ca} –Li, Si, Sr, Ca concentrations in shale sample.

Table 11. Results of leaching samples A1 to A10 rocks using fluids P1 to P13 for Calcium (Ca).

Ca concentrations in samples after leaching of A1 to A10 rocks using fluids P1 to P13

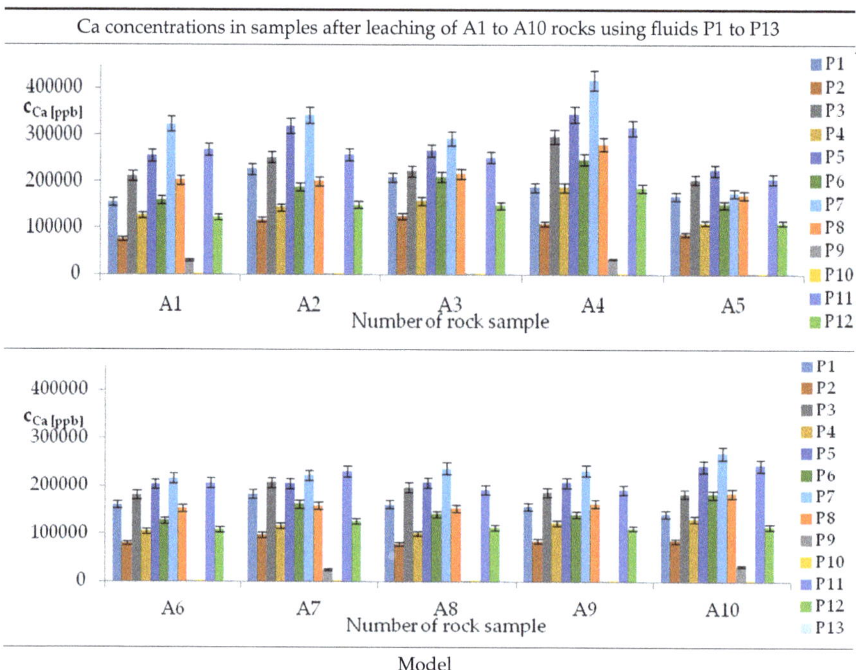

Model

$$c_{Ca} = 1000(3749.56 - 553.95pH - 52.84t + 799.88IS - 83TOC + 39.86TOC^2 + 7.84pH \times t - 322.63IS \times TOC - 1.12G2 - 0.05CS_{Ba} - 0.09CS_{Sr} + 0.000553CS_{Ca} \pm [27.07])$$

$$c'_{Ca} = 1000(-553.95pH - 52.84t + 799.88IS - 83TOC + 39.86TOC^2 + 7.84pH \times t - 322.63IS \times TOC)$$

R^2	0.90	Residual analysis
p	<0.000001	
a	0.080	
α = arctan (a)	4.6°	
Pearson	0.29	
c'_{min} (t = 60 °C)	−3897280	
c'_{max} (t = 60 °C)	−3527670	
c'_{min} (t = 80 °C)	−3836130	
c'_{max} (t = 80 °C)	−3172780	

c_{Ca}—concentration of calcium in leachates; r_A—residual; a-directional coefficient; arctan-arcus tangent; Pearson-Pearson correlation coefficient for r_A and c_B; c'_{min}(t = 60 °C); c'_{min}(t = 80 °C); c'_{max}(t = 60 °C); c'_{max}(t = 80 °C)—explained before Table 5; CS_{Ba}, CS_{Ca}, CS_{Sr}-Ba, Ca, Sr concentrations in shale sample.

Table 12. Results of leaching samples A1 to A10 rocks using fluids P1 to P13 for Magnesium (Mg).

Mg concentrations in samples after leaching of A1 to A10 rocks using fluids P1 to P13

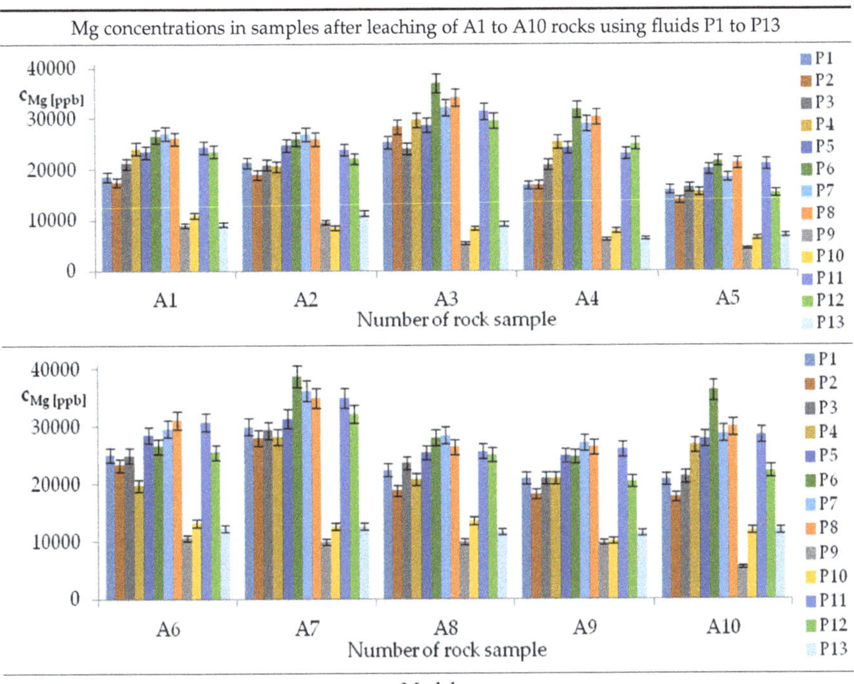

Model

$$c_{Mg} = -16117 - 8518pH + 1756.39t - 21296IS + 35845TOC + 3511.7TOC^2 + 30118pH \times IS - 726.9t \times TOC - 294G2 - 775G3 - 459Aw - 8.2CS_{Ba} + 1.6009CS_{Mg} - 0.1145CS_{Si} \pm [2311]$$

$$c_{Mg} = -8518pH + 1756.39t - 21296IS + 35845TOC + 3511.7TOC^2 + 30118pH \times IS - 726.9t \times TOC$$

R^2	0.92	Residual analysis
p	<0.000001	
a	0.079	
α = arctan (a)	4.5°	
Pearson	0.28	
c'_{min} (t = 60 °C)	26676	
c'_{max} (t = 60 °C)	57888	
c'_{min} (t = 80 °C)	30668	
c'_{max} (t = 80 °C)	92144	

c_{Mg}—concentration of magnesium in leachates; r_A—residual; a-directional coefficient; arctan-arcus tangent; Pearson-Pearson correlation coefficient for r_A and c_B; c'_{min}(t = 60 °C); c'_{min}(t = 80 °C); c'_{max}(t = 60 °C); c'_{max}(t = 80 °C)—explained before Table 5; CS_{Ba}, CS_{Mg}, CS_{Si}-Ba, Mg, Si concentrations in shale sample.

Table 13. Results of leaching samples A1 to A10 rocks using fluids P1 to P13 for Silicon (Si).

Si concentrations in samples after leaching of A1 to A10 rocks using fluids P1 to P13

Model

$$c_{Si} = -298621 + 9229pH - 35067IS - 6556pH^2 + 457TOC^2 - 338pH \times TOC + 324t \times IS - 151.6G1 - 68G5 + 0.0555CS_{Si} \pm [2312]$$

$$c_{Si} = 92297pH - 35067IS - 6556pH^2 + 457TOC^2 - 338pH \times TOC + 324t \times IS$$

R^2	0.96	Residual analysis
p	<0.000001	
a	0.049	
α = arctan (a)	2.8°	
Pearson	0.30	
c'_{min} (t = 60 °C)	286589	
c'_{max} (t = 60 °C)	316728	
c'_{min} (t = 80 °C)	289889	
c'_{max} (t = 80 °C)	320028	

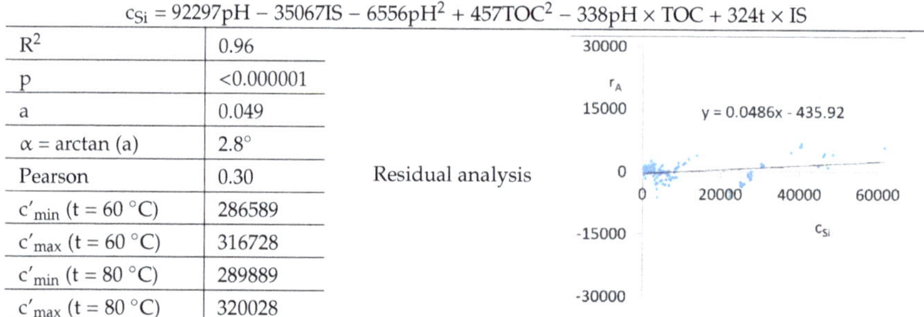

c_{Si}—concentration of silicon in leachates; r_A-residual; a—directional coefficient; arctan-arcus tangent; Pearson-Pearson correlation coefficient for r_A and c_B; c'_{min}(t = 60 °C); c'_{min}(t = 80 °C); c'_{max}(t = 60 °C); c'_{max}(t = 80 °C)—explained before Table 5; CS_B, CS_{Mo}, CS_{Sr} –Ba, Mo, Sr concentrations in shale sample.

To derive general equations, the entire available data set for fluid variables as well as rocks was analyzed. Nine general equations describing the shale rock leaching with model fracturing fluids were obtained. For all equations, the determination coefficient values are in the range from 0.86 to 0.96. This means that the presented model matches are of high significance.

It was assumed that correctly performed regression allows to obtain a graph of residues from real values in the form of points approximately arranged on both sides of a straight line parallel to the abscissa, which represents the actual values. As an approximation, a straight line with the inclination angle of less than 22.5° was drawn, since the position of the points on both sides of the line inclined to abscissa at an angle of 45° means that a sufficient number of variables were not included in the description of the model.

Pearson correlation coefficients between laboratory data and residuals (difference between laboratory and model data) were lower than 0.5 (and higher than 0.5), which means that residuals are poorly correlated with the result. Residual analysis gives information that the applied variables describe the model in the tested range well enough and sufficiently. Moreover, no correlation occurrence between residuals and measured values means that there are probably no methodological errors.

Table 14. Possible range of selected elements decreased in flowback fluid.

Model	Concentration Operating Range (ppb)		Analysed Concentration Range in MFF (ppb)	
	60 °C	80 °C	min	max
B	729	978	371	6044
Ba	1217	1371	25	4219
Sr	1210	1360	125	2687
Mo	626	89	25	760
Rb	101.6	101.6	25	249
Li	44.8	29.8	25	186
Ca	369608	663354	130	417450
Mg	31212	61476	4316	38782
Si	30139	30139	250	64153

For the obtained models it was calculated that appropriate control of the fracturing fluid parameters can significantly reduce the amount of leached contaminants (Table 14). The maximum value of possible changes in fixed temperature was named Concentration Operating Range (abr. COR). Concentration range (min/max) in MFF (min/max) are given as minimum and maximum values of components concentrations in analysed fluids after leaching tests. For boron, the possibility of lowering the concentration in flowback fluid is over 10 to almost 15% at 80 °C, while for lithium it is from 16 to 24% and for barium over 25%. Considerably greater COR was calculated for strontium, rubidium and silica, about 50%, for calcium and magnesium, so it is theoretically possible to select the parameters of the fluid giving no leaching of a given element. The biggest dependence on temperature was observed for molybdenum where at 60 °C the concentration could be increased by about 10% whereas at 80 °C by 80%. This is a direct argument that such a model development approach is justified and has a potential for practical use. Similar research needs to be continued and more advanced models should be developed according to scheme presented in Figure 3. Such models using data from many different reservoirs, evaluated and improved using field data, could have important impact on designing the flowback treatment technologies immediately after drilling. However using only fast and easy procedures of rock analysis is crucial for this approach.

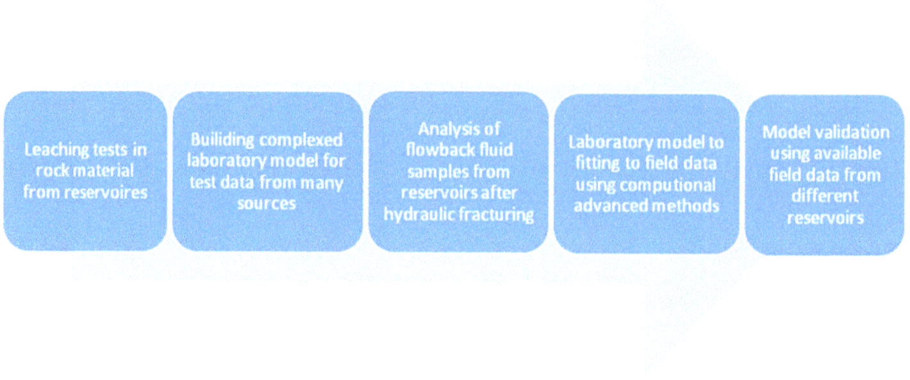

Figure 3. Possible directions of further studies on the development of proposed model.

4. Conclusions

The results of mineralogical and elemental analysis as well as the determination of organic carbon in the studied shale rock samples from various sections of the horizontal section of the exploration well, indicate high heterogeneity of the deposit at a macroscopic scale. A high divergence in the composition made it possible to make a model with a wide range of applications. Unfortunately, in the shale systems, a high heterogeneity of the mineral and elemental composition along the created fractures may occur. This can strongly limit the presented and other models and always has to be considered before its application as a developed industry tool. Nevertheless, using cross-sectional samples from whole fractured parts of the wellbore deliver the most technical/economical possible approximations of the shale composition in a chosen part of the reservoir.

For the model, fracturing fluid-shale rock, mathematical relationships based on T, ionic strength, TOC, pH of fracturing fluid and elemental composition, mineralogical composition, and surface area of shale rock were derived. This approach allows one to predict the concentration of selected metals in a wide range of variables. Nine equations were obtained, the coefficient of determination over 0.85 was obtained for the whole set with confidence level above 0.99. For chosen model the standard deviation was always the lowest among other taken into consideration and its value justifies the correctness of application of R^2 as a crucial parameter to choose most adequate model. Residual analysis gives information that applied variables describes the model in the tested range well enough. Moreover, no correlation between residuals and measured values means that there are probably no methodological errors. Selected parameters of the model fracturing fluid and rock properties are sufficient to predict the composition of the flowback fluid in the laboratory system. In the next research other shale rock samples should be tested and data from analysis of flowback fluid for attempts to develop industrial version of the model need to be carried.

Unfortunately, a comparison with other models in the literature is not possible due to unnormalized character off the data and a lack of model description from other sources. However, the presented paper may become a significant contribution in the modelling development approach in the shales leaching systems. Test results can be a good basis for developing a commercial tool for predicting the treatment fluid at the design stage of hydraulic fracturing.

Author Contributions: Conceptualization, A.R.; Data curation, A.R.; Formal analysis, A.R.; Funding acquisition, A.R.; Investigation, A.R.; Methodology, A.R.; Project administration, J.H.; Resources, J.H.; Software, A.R.; Validation, K.K. and J.H.; Visualization, A.R.; Writing-original draft, A.R.; Writing-review & editing, K.K. and J.H.

Funding: This research received no external funding.

Conflicts of Interest: The authors declare no conflict of interest.

Appendix A

Table A1. Detailed results of XRD analysis of shales samples.

No. Shale Sample	Mineral (%)														
	Quarz	Alb	Micr	Ort	Calc	Dol	Pir	Bar	Chl	Ill	Mon	Kao	Mus	Bio	Par	Gla
A1	15.89	2.93	1.58	1.10	9.38	2.56	6.70	2.43	4.75	1.10	0.00	9.38	38.29	0.36	2.56	0.98
A2	21.58	1.83	1.10	0.74	2.93	0.74	0.00	0.08	7.19	0.74	0.00	2.81	57.48	1.58	0.61	0.61
A3	12.74	2.07	2.68	0.61	28.67	8.53	0.00	0.12	3.77	0.48	10.73	5.85	18.64	2.93	1.83	0.36
A4	19.58	3.10	5.08	0.37	21.52	3.97	0.00	0.25	7.81	8.80	0.00	1.86	2.61	22.81	1.11	1.11
A5	23.73	0.98	0.86	0.37	22.81	4.80	0.00	0.12	9.34	2.83	0.12	11.55	19.17	0.62	0.86	1.85
A6	22.57	3.44	2.59	0.74	13.05	30.67	1.73	0.12	4.06	1.73	0.00	0.37	13.29	0.24	3.32	2.09
A7	13.60	2.70	3.19	0.12	12.86	33.97	0.12	0.12	5.03	3.19	0.24	0.36	17.27	4.53	0.50	2.20
A8	25.66	4.16	2.08	0.62	8.44	16.50	0.00	0.12	4.88	0.74	0.00	0.36	33.76	0.48	1.72	0.48
A9	34.97	6.67	3.52	1.00	9.17	11.19	0.76	0.12	5.04	2.13	0.00	2.13	5.53	3.77	1.51	12.47
A10	21.81	2.18	1.09	0.48	9.82	12.13	0.12	0.00	7.52	0.48	0.00	1.21	41.45	0.36	0.97	0.36

Alb—albite, Micr—microclase, Ort—orthoclase, Calc—calcite, Dol—dolomite, Pir—pyrite, Bar-barite, Chl—chlorite, Ill—illite, Mon—montmorillonite, Kao—kalonite, Mus—muscovite, Bio—biotite, Par—paragonite, Gla—glaukonite

References

1. Dayal, A.M. Shale. In *Shale Gas: Exploration and Environmental and Economic Impacts*; Elsevier Science: Amsterdam, Holand, 2017; ISBN 9780128095355.
2. Singh, K.; Holditch, S.A.; Ayers, W.B. Basin Analog Investigations Answer Characterization Challenges of Unconventional Gas Potential in Frontier Basins. *J. Energy Resour. Technol.* **2008**, *130*. [CrossRef]
3. Chopra, S.; Solutions, A.S.; Kumar, R.; Marfurt, K.J. Current Workflows for Shale Gas Reservoir Characterization. In Proceedings of the Unconventional Resources Technology Conference, Denver, CO, USA, 12–14 August 2013. [CrossRef]
4. Chermak, J.A.; Schreiber, M.E. Mineralogy and trace element geochemistry of gas shales in the United States: Environmental implications. *Int. J. Coal Geol.* **2014**, *126*, 32–44. [CrossRef]
5. Liu, J.; Yao, Y.; Elsworth, D.; Liu, D.; Cai, Y.; Dong, L. Vertical heterogeneity of the shale reservoir in the lower silurian longmaxi formation: Analogy between the southeastern and Northeastern Sichuan Basin, SW China. *Minerals* **2017**, *7*, 151. [CrossRef]
6. Barnhoorn, A.; Houben, M.E.; Lie-A-Fat, J.; Ravestein, T.; Drury, M. Variations in petrophysical properties of shales along a stratigraphic section in the Whitby mudstone (UK). In Proceedings of the EGU General Assembly 2015, Vienna, Austria, 12–17 April 2015.
7. Chen, L.; Lu, Y.; Jiang, S.; Li, J.; Guo, T.; Luo, C. Heterogeneity of the lower silurian longmaxi marine shale in the southeast sichuan basin of China. *Mar. Pet. Geol.* **2015**, *65*, 232–246. [CrossRef]
8. Rogala, A.; Krzysiek, J.; Bernaciak, M.; Hupka, J. Non-aqueous fracturing technologies for shale gas recovery. *Physicochem. Probl. Miner. Process.* **2013**, *49*, 313–322.
9. Rogala, A.; Ksiezniak, K.; Krzysiek, J.; Hupka, J. Carbon dioxide sequestration during shale gas recovery. *Physicochem. Probl. Miner. Process.* **2014**, *50*, 681–692.
10. Howard, G.C.; FAST, C.R. *Hydraulic Fracturing*; Henry L. Doherty Memorial Fund of AIME: New York, NY, USA, 1970; Volume 2, ISBN 0895202018.
11. Tao, H.; Zhang, L.; Liu, Q.; Deng, Q.; Luo, M.; Zhao, Y. An Analytical Flow Model for Heterogeneous Multi-Fractured Systems in Shale Gas Reservoirs. *Energies* **2018**, *11*, 3422. [CrossRef]
12. Economides, M.J.; Martin, T. *Modern Fracturing—Enhancing Natural Gas Production*; Energy Tribune Publishing: Houston, TX, USA, 2007; 509p.
13. Gandossi, L. *An Overview of Hydraulic Fracturing and Other Formation Stimulation Technologies for Shale Gas Production*; Joint Research Centre: Mercier, Luxembourg, 2015; ISBN 9789279347290.
14. Mader, D. *Hydraulic Proppant Fracturing and Gravel Packing*; Elsevier Science: Amsterdam, Holand, 1989; ISBN 9780444873521.
15. Ksiezniak, K.; Rogala, A.; Hupka, J. Wettability of shale rock as an indicator of fracturing fluid composition. *Physicochem. Probl. Miner. Process.* **2015**, *51*, 315–323.
16. Albrycht, I.; Boy, K.; Jankowski, J.M. *Gaz Niekonwencjonalny—Szansa dla Polski i Europy? Analiza i Rekomendacje*; Instytut Kościuszki: Kraków, Poland, 2011; ISBN 9788393109340.
17. Arthur, J.D.; Bohm, B.K.; Cornue, D. Environmental Considerations of Modern Shale Gas Development. In Proceedings of the SPE Annual Technical Conference and Exhibition, San Antonio, TX, USA, 9–11 October 2009.
18. Hayes, T.; Severin, B.F. *Barnett and Appalachian Shale Water Management and Reuse Technologies*; Project Report by Gas Technology Institute Research Partners toSecure Energy for America; Publications Office of the European Union: Mercier, Luxembourg, 2012; pp. 1–10.
19. Boschee, P. Produced and Flowback Water Recycling and Reuse: Economics, Limitations, and Technology. *Oil Gas Facil.* **2014**, *3*, 16–21. [CrossRef]
20. Abualfaraj, N.; Gurian, P.L.; Olson, M.S. Assessing residential exposure risk from spills of flowback water from Marcellus shale hydraulic fracturing activity. *Int. J. Environ. Res. Public Health* **2018**, *11*, 15. [CrossRef]
21. Zhou, J.; Zhang, L.; Braun, A.; Han, Z. Investigation of processes of interaction between hydraulic and natural fractures by PFC modeling comparing against laboratory experiments and analytical models. *Energies* **2017**, *10*, 1001. [CrossRef]
22. Clarkson, C.R.; Williams-Kovacs, J. Modeling Two-Phase Flowback of Multifractured Horizontal Wells Completed in Shale. *SPE J.* **2013**, *18*. [CrossRef]

23. Williams-Kovacs, J.D.; Clarkson, C.R. A modified approach for modeling two-phase flowback from multi-fractured horizontal shale gas wells. *J. Nat. Gas Sci. Eng.* **2016**, *30*, 127–147. [CrossRef]
24. Clarkson, C.R.; Haghshenas, B.; Ghanizadeh, A.; Qanbari, F.; Williams-Kovacs, J.D.; Riazi, N.; Debuhr, C.; Deglint, H.J. Nanopores to megafractures: Current challenges and methods for shale gas reservoir and hydraulic fracture characterization. *J. Nat. Gas Sci. Eng.* **2016**. [CrossRef]
25. Jia, P.; Cheng, L.; Clarkson, C.R.; Huang, S.; Wu, Y.; Williams-Kovacs, J.D. A novel method for interpreting water data during flowback and early-time production of multi-fractured horizontal wells in shale reservoirs. *Int. J. Coal Geol.* **2018**, *200*, 186–196. [CrossRef]
26. Cao, P.; Liu, J.; Leong, Y.K. A multiscale-multiphase simulation model for the evaluation of shale gas recovery coupled the effect of water flowback. *Fuel* **2017**, *199*, 191–205. [CrossRef]
27. Bai, B.; Elgmati, M.; Zhang, H.; Wei, M. Rock characterization of Fayetteville shale gas plays. *Fuel* **2013**, *105*, 642–652. [CrossRef]
28. Zhang, H.Y.; Gu, D.H.; Zhu, M.; He, S.L.; Men, C.Q.; Luan, G.H.; Mo, S.Y. Optimization of Fracturing Fluid Flowback Based on Fluid Mechanics for Multilayer Fractured Tight Reservoir. *Adv. Mater. Res.* **2014**, *886*, 448–451. [CrossRef]
29. Michel, G.; Civan, F.; Sigal, R.; Devegowda, D. Proper Simulation of Fracturing-Fluid Flowback from Hydraulically-Fractured Shale-Gas Wells Delayed by Non-Equilibrium Capillary Effects. In Proceedings of the Unconventional Resources Technology Conference, Denver, CO, USA, 12–14 August 2013.
30. Moray, L.; Holdaway, K.R. Fluid flowback prediction. U.S. Patent US20150112597A1, 23 April 2015.
31. Jurus, W.J.; Whitson, C.H.; Golan, M. Modeling Water Flow in Hydraulically-Fractured Shale Wells. In Proceedings of the SPE Annual Technical Conference and Exhibition, New Orleans, LA, USA, 30 September–2 October 2013.
32. Abdulelah, H.; Mahmood, S.; Al-Hajri, S.; Hakimi, M.; Padmanabhan, E. Retention of Hydraulic Fracturing Water in Shale: The Influence of Anionic Surfactant. *Energies* **2018**, *11*, 3342. [CrossRef]
33. He, C.; Li, M.; Liu, W.; Barbot, E.; Vidic, R.D. Kinetics and Equilibrium of Barium and Strontium Sulfate Formation in Marcellus Shale Flowback Water. *J. Environ. Eng.* **2014**, *140*. [CrossRef]
34. Barbot, E.; Vidic, N.S.; Gregory, K.B.; Vidic, R.D. Spatial and temporal correlation of water quality parameters of produced waters from Devonian-age shale following hydraulic fracturing. *Environ. Sci. Technol.* **2013**, *47*, 2562–2569. [CrossRef] [PubMed]
35. Gdanski, R.; Weaver, J.; Slabaugh, B. A New Model for Matching Fracturing Fluid Flowback Composition. In Proceedings of the SPE Hydraulic Fracturing Technology Conference, College Station, TX, USA, 29–31 January 2007.
36. Liu, X.; Ortoleva, P. A General-Purpose, Geochemical Reservoir Simulator. *Soc. Pet. Eng.* **1996**. [CrossRef]
37. Balashov, V.N.; Engelder, T.; Gu, X.; Fantle, M.S.; Brantley, S.L. A model describing flowback chemistry changes with time after Marcellus Shale hydraulic fracturing. *Am. Assoc. Pet. Geol. Bull.* **2015**, *99*, 143–154. [CrossRef]
38. Kalbe, U.; Berger, W.; Eckardt, J.; Simon, F.G. Evaluation of leaching and extraction procedures for soil and waste. *Waste Manag.* **2008**, *28*, 1027–1038. [CrossRef] [PubMed]
39. Fällman, A.M.; Aurell, B. Leaching tests for environmental assessment of inorganic substances in wastes, Sweden. *Sci. Total Environ.* **1996**, *178*, 71–84. [CrossRef]
40. Mahmoudkhani, M.; Wilewska-Bien, M.; Steenari, B.M.; Theliander, H. Evaluating two test methods used for characterizing leaching properties. *Waste Manag.* **2008**, *28*, 133–141. [CrossRef]
41. Quevauviller, P.; Van der Sloot, H.A.; Ure, A.; Muntau, H.; Gomez, A.; Rauret, G. Conclusions of the workshop: Harmonization of leaching/extraction tests for environmental risk assessment. *Sci. Total Environ.* **1996**, *178*, 133–139. [CrossRef]
42. Organization for Economic Cooperation and Development. *OECD Guidelines for Testing of Chemicals*; OECD Publishing: Paris, France, 2000; ISBN 9108026995001.
43. RStudio Team. *RStudio: Integrated Development for R*; RStudio, Inc.: Boston, MA, USA, 2015. Available online: http://www.rstudio.com/ (accessed on 1 April 2019).

© 2019 by the authors. Licensee MDPI, Basel, Switzerland. This article is an open access article distributed under the terms and conditions of the Creative Commons Attribution (CC BY) license (http://creativecommons.org/licenses/by/4.0/).

Article

Experimental Investigation of the Effects of Drilling Fluid Activity on the Hydration Behavior of Shale Reservoirs in Northwestern Hunan, China

Han Cao [1,2,3], Zheng Zhang [1,2,*], Ting Bao [4,*], Pinghe Sun [1,2,*], Tianyi Wang [1,2] and Qiang Gao [1,2]

1 Schools of Geosciences and Info-Physics, Central South University, Changsha 410083, China
2 Key Laboratory of Metallogenic Prediction of Nonferrous Metals and Geological Environment Monitoring (Ministry of Education), Central South University, Changsha 410083, China
3 Key Laboratory of Oil and Gas Reservoir Geology and Exploitation, Southwest Petroleum University, Chengdu 610500, China
4 Department of Civil and Environmental Engineering, Michigan Technological University, 1400 Townsend Drive, Houghton, MI 49931, USA
* Correspondence: 175011077@csu.edu.cn (Z.Z.); tbao@mtu.edu (T.B.); pinghesun@csu.edu.cn (P.S.); Tel.: +86-133-1956-9196 (H.C.)

Received: 25 May 2019; Accepted: 13 August 2019; Published: 16 August 2019

Abstract: The interaction between drilling fluid and shale has a significant impact on wellbore stability during shale oil and gas drilling operations. This paper investigates the effects of the drilling fluid activity on the surface and osmotic hydration characteristics of shale. Experiments were conducted to measure the influence of drilling fluid activity on surface wettability by monitoring the evolution of fluid-shale contact angles. The relationship between drilling fluid activity and shale swelling ratio was determined to investigate the osmotic hydration behavior. The results indicate that, with increasing drilling fluid activity, the fluid–shale contact angles gradually increase—the higher the activity, the faster the adsorption rate; and the stronger the inhibition ability, the weaker the surface hydration action. The surface adsorption rate of the shale with a KCl drilling fluid was found to be the highest. Regarding the osmotic hydration action on the shale, the negative extreme swelling ratio (b) of the shale was found to be: $b_{KCl} < b_{CTAB} < b_{SDBS}$. Moreover, based on the relationship between the shale swelling ratio and drilling fluid activity, shale hydration can be divided into complete dehydration, weak dehydration, surface hydration, and osmotic hydration, which contributes to the choice of drilling fluids to improve wellbore stability.

Keywords: shale reservoirs; surface hydration; osmotic hydration; negative extreme swelling ratio; safety levels of activity

---

1. Introduction

Wellbore instability is a worldwide challenge in the oil drilling industry [1], causing direct economic losses of more than 1 billion USD annually [2,3]. It has been estimated that about 75% of drilled strata are shale formations and that 90% of wellbore instability occurs in such formations [3–5]. Shale is a sedimentary rock with distinct laminated layers and a high clay content [5,6]. When drilling into a shale formation, the shale comes into contact with the water-based drilling fluid. Thereby, the water molecules in the drilling fluid are adsorbed by the inter-layer of the clay minerals and onto the surface of the clay particles by coordination, electrostatic interaction, and hydrogen bonding;

and surface hydration occurs [3,7]. Additionally, due to the presence of micro-cracks in the shale, free water can easily enter the interior of the rock and induce a series of physico-chemical and mechanical changes [8]. The mechanical changes promote the expansion and enlargement of the original micro-cracks inside the shale, and then macro-cracks form, providing channels for the water molecules to enter the shale, which increases the contact area between fluid and clay particles and consequently leads to ion hydration and osmotic hydration [9]. Shale hydration is responsible for the attenuation of inter-particle interaction and cementation, which result in a decrease in the compressive strength and hardness of the rock [3,10]. Al-Bazali et al. [11] confirmed that diffusion osmosis has a detrimental effect on the mechanical stability of shale by reducing its compressive strength. As the degree of hydration increases, the rock structure is further destroyed, causing the wellbore to expand and fall off, resulting in wellbore instability. Van Oort et al. [12] found that shale–fluid interactions could be controlled to decrease shale hydration, which can enhance wellbore stabilization.

The factors that cause shale hydration are complex: pressure differentials, chemical potential differences [12], the crack structure of shale, and the type and content of clay mineral [13]. Additionally, the difference between the activity of the drilling fluid and the activity of shale formation has been proven to be a crucial factor to affect shale hydration [14]. Al-Bazali et al. [15] found that the membrane efficiency decided by the difference in activity between the drilling fluid and the shale formation was related to the types of cation and anion in water-based fluids, which affect shale hydration. Chenevert et al. [16] proposed the equilibrium activity theory of drilling fluids, and found that the higher the water activity in an oil-based drilling fluid, the more severe the shale hydration. The authors concluded the following: (1) when the shale formation activity is less than the drilling fluid activity, the water molecules in the drilling fluid can flow into the shale formation. Thus, surface hydration or osmotic hydration of the shale formation can be induced, which results in shale inflation; (2) when the shale formation activity is greater than the drilling fluid activity, fluid in the shale formation migrates into the drilling fluid. Thus, dehydration occurs, which induces shale contraction; and (3) when the activity of the drilling fluid is the same as that of the shale, there is no water exchange between the drilling fluid and the shale formation, and the rock remains in its original state [13,17,18]. Therefore, investigating the impact of drilling fluid activity on shale hydration behaviors has practical significance for improving mining efficiency and wellbore stability.

Shale hydration is divided into three stages: surface hydration, ion hydration, and osmotic hydration [19]. In previous studies, only the surface hydration of shale was considered as an important factor related to the surface wettability of shale, and was mainly related with fluid adsorption characterized by the change in the contact angle of the fluid on the surface of the shale; the faster the fluid adsorption rate, the faster the rate of change of the contact angle. Meanwhile, osmotic hydration was considered to be a factor inducing the swelling of shale, and was usually characterized by the swelling ratio of shale [20,21]. Huang et al. [19] found that a surfactant compound of polyamine (PA) and twelve alkyl two hydroxyethyl amine oxide (THAO) could enhance the surface hydrophobicity after adsorption onto the surface of shale particles and could restrain osmotic hydration through swapping out inorganic cations in the clay inter-layer. Yue et al. [5] found that the absorption of the cetyltrimethylammonium bromide (CTAB) cation on the surface of shale could increase the contact angle between the drilling fluid and the shale and inhibit the surface hydration of shale. Cai et al. [22] found that a composite surfactant could effectively change the wettability performance of water-based drilling fluid; this allowed the stability of the shale to be enhanced by controlling the wettability of the drilling fluid. Aghil Moslemizadeh et al. [23] reported the effect of silica nanoparticles (NPs) as a physical sealing agent on water invasion into the Kazhdumi shale; the authors showed that the use of the NPs reduced shale hydration and improved the wellbore stability. Liu et al. [24] concluded that when shale comes into contact with water-based drilling fluid, surface hydration occurs on the surface of the shale clay particles, thus forming a surface hydration film, generating hydration stress, destroying the original mechanical balance, and eventually causing the shale strength to decrease. Using shale osmotic hydration experiments based on the generalized Usher model, Wen et al. [17] established a model related to the activity, swelling ratio, and hydration degree of shale, and classified shale hydration into three stages: dehydration, surface hydration, and osmotic hydration. However, in the above studies, surface hydration and osmotic hydration were not comprehensively considered, nor were they in studies aimed at investigating the relationship between the difference in the activity of the drilling fluid and the shale and shale hydration. Moreover, the relationship between the safety level of fluid activity and shale hydration has not been established. Determining this relationship could allow the impact of fluid activity on shale hydration to be characterized.

The present work takes advantage of the fact that, since the radius of K^+ is similar to that of the hexagonal cavity in the clay crystal unit, CTA^+ can be embedded into the inter-layer space of montmorillonite and SO_3^- can be adsorbed by polar substances on the surface of shale, which can prevent shale hydration. Therefore, potassium chloride (KCl), sodium dodecylbenzene sulfonate (SDBS), and CTAB drilling fluid systems, as typical inorganic and organic drilling fluids, are widely adopted to maintain wellbore stability [25–27]. Thus, in this study, three drilling fluid systems with different activities were selected as test fluids to investigate the effects of the activity of drilling fluids on the surface hydration and osmotic hydration of shale using tests of the contact angle between the drilling fluid and shale and swelling ratio tests. Furthermore, an evolution mechanism of fluid–shale contact angles as a function of fluid activity was proposed. Then, the safety of drilling fluid activity was classified into four levels based on the relationship between the shale swelling ratio and fluid activity.

2. Materials and Methods

2.1. Materials

In this study, the cationic surfactant used was CTAB, the anionic surfactant used was SDBS, and the inorganic salt used was KCl. These materials were obtained from a chemical company in Tianjin, China. The composition of the KCl, SDBS, and CTAB drilling fluid systems is shown in Table 1.

Table 1. Compositions of drilling fluid systems.

Based Fluid Formula	Inhibitor Concentration
0.35% CMC + 0.08% PH	0.5%~2.5% KCl
	0.5%~2.5% SDBS
	0.5%~2.5% CTAB

Note: CMC is carboxymethylcellulose sodium; PH is potassium humate.

The shale samples used in this study (Figure 1a) were collected from the Longshan shale outcrop (Figure 1b). In order to minimize the influence of shale anisotropy on the experimental results, the shale samples were taken from the same outcrop and drilled perpendicular to the bedding direction. The samples were sanded into 15 shale samples with a diameter of ϕ50 mm and thickness of 5 mm. Then, the samples were dried and kept at 40 °C for 24 hours in an incubator for fluid–shale contact angle tests. Furthermore, 45 shale samples with a diameter of ϕ25 mm and a thickness of 20 mm were prepared for swelling ratio tests. The compositions (Table 2) of the samples indicate that a large amount of brittle minerals and montmorillonite are present in this formation, with montmorillonite comprising 20~27%. The physical and mechanical parameters of the samples are shown in Table 3.

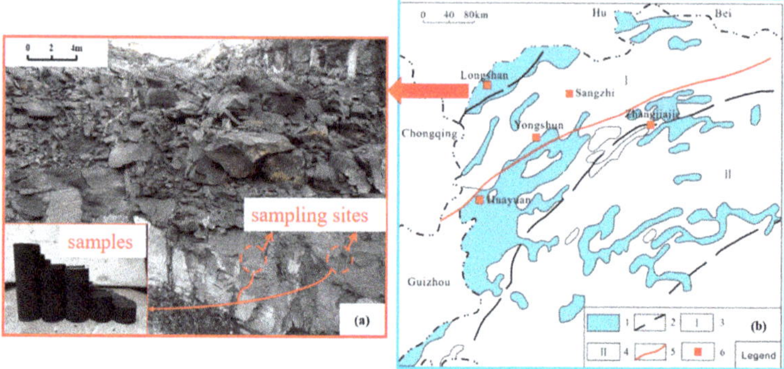

Figure 1. A map of the regional structure and shale outcrop of the Lower Cambrian Niutitang Formation in northwestern Hunan Province, modified after Cao et al., 2018 [28]. 1: Outcrop distribution area; 2: Stratigraphic boundary; 3: Northwestern Hunan area; 4: Xuefeng District, central Hunan; 5: Deep fault; 6: Sampling sites.

Table 2. Contents of mineral components in shale samples.

Quartz (%)	Feldspar (%)	Clay Mineral (%)	Anatase (%)	MICA (%)	Amorphous (%)
21~30	10~15	20~27	5~8	4~9	21~29

Table 3. Physical and mechanical parameters of shale samples.

Elastic Modulus (GPa)	Poisson's Ratio	Brittleness Index	Porosity (%)	Cohesion (MPa)	Internal Friction Angle (°)
1.25~3.86	0.19~0.24	0.20~0.37	2.24~4.47	0.57~0.98	33.82~41.70

2.2. Methods

In this study, tests were conducted for drilling fluid activity, fluid–shale contact angle, and shale swelling ratio.

2.2.1. Fluid Activity Tests

Place 10 ml of the prepared drilling fluid in a glass container, place it in the test box of a calibrated AW-1 water activity meter (Wuxi Huake Instrument and Meter Company, Wuxi, China; measurement range: 0~0.999; precision: 0.01), and then start the test at 23 °C. When the vapor pressure of the sample is the same as that of the surrounding air, the water vapor pressure in the gas space is taken as the value of the sample vapor pressure. Then, the water activity of the drilling fluid is calculated based on the relationship between the water activity and the vapor pressure.

2.2.2. Fluid–Shale Contact Angle Tests

The prepared shale sample was placed on the sample stage of a JCY contact angle meter (Shanghai Fangrui Instrument Co., Ltd., Shanghai, China; contact angle resolution of 0.01°). Then, 5 μl of drilling fluid was dripped onto the shale sample using a micro-syringe. The experimental process was recorded using a MV-1300UM-MIN high-speed camera (Weishi Digital Image Technology Co., Ltd., Beijing, China) with a frame length of 10 ms. Finally, the contact angle of the drilling fluid on the shale surface was obtained using the CONTA3.0 contact angle analysis software (Shanghai Fangrui Instrument Co., Ltd., Shanghai, China). To reduce experimental error, the contact angle of the drilling fluid was measured at three different locations on the shale surface and the average value was taken as the final measurement result.

2.2.3. Shale Swelling Ratio Tests

A displacement sensor (Si Mingwei Technology Co., Ltd., Shenzhen, China) was attached to the prepared shale sample, and the sample was then hung on a pylon. Then, 100 mL of drilling fluid was placed in a liquid cup of a CPZ-II swelling meter (Qingdao Jiaonan Analytical Instrument Factory, Qingdao, China) and the shale sample was immersed in the liquid for eight hours. The swelling meter records the amount of shale swelling displacement in real time. In order to improve the accuracy of the experimental results, three shale samples were used in three replicate experiments for each drilling fluid, and the average swelling ratio was taken as the final measurement result.

3. Results and Discussion

3.1. Effect of Inhibitor Concentration on Drilling Fluid Activity

The relationship between the inhibitor concentration and drilling fluid activity is shown in Table 4.

Table 4. The relationship between inhibitor concentration and drilling fluid activity.

Inhibitor Concentration		0.5%	1.0%	1.5%	2.0%	2.5%
KCl	drilling fluid activities	0.860	0.871	0.900	0.933	0.957
SDBS		0.904	0.922	0.930	0.935	0.946
CTAB		0.911	0.928	0.952	0.970	0.988

As the concentration of inhibitors in the drilling fluid system increases, the activity of the drilling fluid shows an increasing trend. At the same concentration, the CTAB drilling fluid has the highest activity and the KCl drilling fluid has the lowest activity. This means that at the same concentration, the difference in activity between the shale and the drilling fluid is greatest for the KCl drilling fluid, so when the shale interacts with the drilling fluid, the driving force pushing the water molecules in the drilling fluid into the shale is greatest.

3.2. Effect of Drilling Fluid Activity on Surface Hydration of Shale

The relationship between the drilling fluid activity and the fluid–shale contact angle is shown in Figures 2–4.

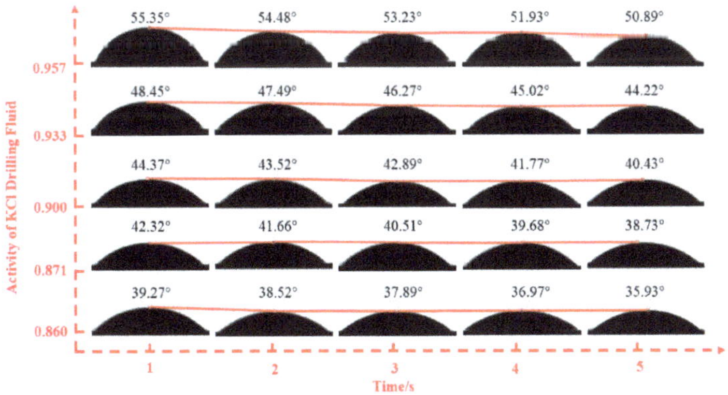

Figure 2. The evolution of fluid–shale contact angles with different activities of KCl drilling fluid.

As shown in Figure 2, as the activity of the drilling fluid increases, the fluid–shale contact angles tend to increase overall. This is due to the fact that the adsorption energy of exchangeable cationic K^+

on the surface of montmorillonite, which has a negative charge, is stronger than the adsorption energy of water molecules on the surface of montmorillonite. Additionally, the exchangeable K^+ is adsorbed on the surface of montmorillonite, forming an ionic film, which greatly improves the hydrophobicity of the surface of shale, which shows strong inhibitory ability of surface hydration. Furthermore, the rate of change of the fluid–shale contact angles decreases as the drilling fluid activity increases, which indicates that with increasing drilling fluid activity, the migration and adsorption rate of K^+ in the fluid increases and the number of layers of ionic film also increases, which can form a dense ionic film that prevents water from entering the shale. This means that the inhibitory effect of the drilling fluid on shale surface hydration is significantly improved when using KCl drilling fluid.

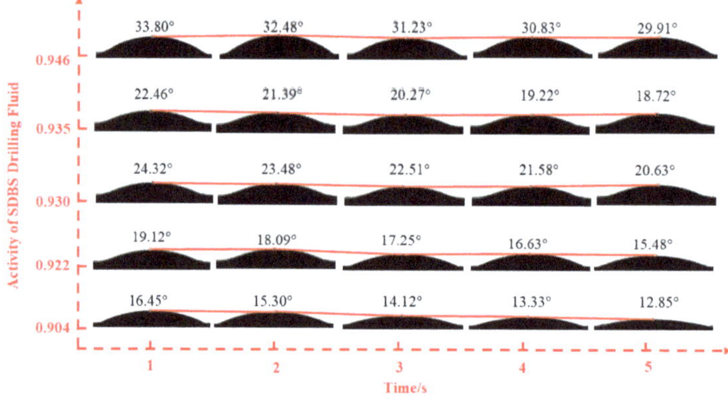

Figure 3. The evolution of fluid–shale contact angles for different activities of sodium dodecylbenzene sulfonate (SDBS) drilling fluid.

The evolution of the fluid–shale contact angle with increasing drilling fluid activity for SDBS drilling fluid is shown in Figure 3. The hydrophilic groups of SDBS molecules can form molecular films on the surface of shale through the van der Waals force due to the large amount of hydrophilic minerals (such as quartz, mica, and montmorillonite) that are present in shale, and which can reduce the fluid–shale interface tension and increase the fluid–shale contact angles. Meanwhile, the polar group SO_3^- in the drilling fluid is adsorbed by polar substances through the van der Waals force, and the SO_3^- can thus form an ionic film on the surface of the shale. As shown in Figure 3, with increasing drilling fluid activity, the adsorption rate of SO_3^- by the difference between the activity of the drilling fluid and the activity of the shale formation is greatly increased, the rate of change of fluid–shale contact angle decreases, and then the thickness of the adsorption layer increases accordingly, improving the hydrophobicity of the shale surface and thus effectively inhibiting the hydration of the shale surface by the fluid.

As can be seen from Figure 4, for the CTAB drilling fluid, the fluid–shale contact angles increase with increasing drilling fluid activity. The hydrophilic groups of the CTAB molecule are adsorbed on the surface of hydrophilic minerals (such as quartz, mica, and montmorillonite) present in the shale via Van der Waals forces, with the hydrophobic end of the molecule facing outward. Meanwhile, the CTA^+ cations in the drilling fluid are adsorbed by O^- anions on the surface of the shale via ionic bonds (Figure 5a). As more surfactant molecules and CTA^+ are adsorbed on the surface of the shale, the thickness and density of the adsorption layer increase, which enhances the hydrophobicity of the shale surface and effectively suppresses the surface hydration of the shale (Figure 5b). As shown in Figure 4, increasing drilling fluid activity accelerates the adsorption rate of CTA^+ and promotes the formation of a hydration-inhibiting layer, which causes the fluid–shale contact angles to increase. The adsorption mechanism of CTAB on the shale surface is illustrated in Figure 5.

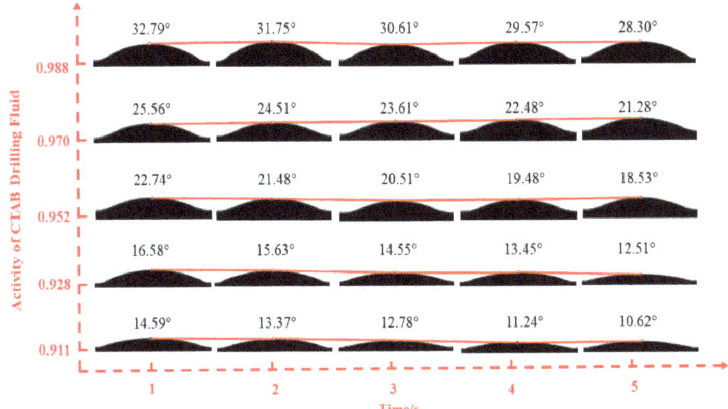

Figure 4. The evolution of fluid–shale contact angle for different activities of cetyltrimethylammonium bromide (CTAB) drilling fluid.

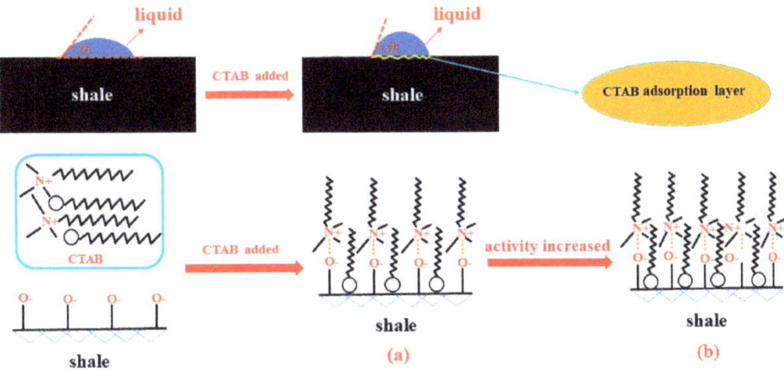

Figure 5. The adsorption mechanism of CTAB on the surface of shale.

The relationship between the drilling fluid activities (α_w) and the change ratio of the fluid–shale contact angle ($\Delta\beta/t$) is shown in Figure 6. It can be seen that with increasing drilling fluid activity the rate of change of the contact angles undergoes a decreasing trend. Increasing drilling fluid activity accelerates the movement of ions or molecules in the drilling fluid, and consequently the rate of adsorption on the shale surface is increased. This effect resulted in the observed reduction of the fluid–shale contact angles in the three drilling fluids over the same period of time. The rate of change of contact angles is the lowest using the KCl drilling fluid, which indicates that the hydration-inhibiting ability of the KCl drilling fluid is the strongest among the three drilling fluids.

3.3. *Effect of Drilling Fluid Activity on the Osmotic Hydration of Shale*

The relationship between the drilling fluid activity and the shale swelling ratio is shown in Figure 7.

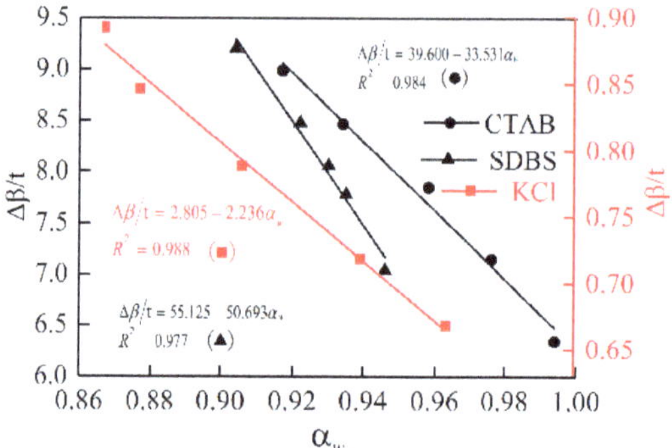

Figure 6. Relationship between drilling fluid activity and the rate of change of fluid–shale contact angle.

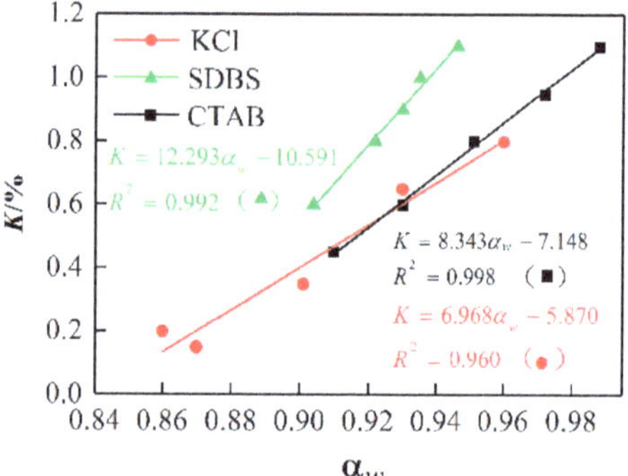

Figure 7. Relationship between drilling fluid activity and shale swelling ratio.

The results demonstrate that the shale swelling ratio increases with increasing drilling fluid activity. As the drilling fluid activity increases, the difference in activity between the drilling fluid and the shale formation increases, meaning that the driving force driving the free water in the drilling fluid into the shale formation increases. Additionally, shale formations contain abundant bedding planes and micro-cracks, which allow drilling fluids to migrate into the formation [29]. The relationship between the drilling fluid activity and the shale swelling ratio can be established by Equation (1):

$$K = a\alpha_w - b \tag{1}$$

where a is the shale swelling ratio per unit activity, which is the slope of the line in Figure 9; and b is the shale swelling ratio when the drilling fluid activity is zero; the swelling ratio, defined as the negative extreme swelling ratio, is equal to the longitudinal intercept of the line. Their units are both 1.

It can also be seen from Figure 7 that there is a significant linear relationship between the activity of the CTAB drilling fluid and the shale swelling ratio, and that there is also a significant linear relationship between the activities of the SDBS and KCl drilling fluids and the swelling ratio. Additionally, when the swelling ratio is close to zero, the drilling fluid activity can be defined as the critical drilling fluid activity (α_{wo}), whose limit is equal to the shale formation activity (α_s). When the drilling fluid activity is lower than the shale formation activity, dehydration of the shale occurs due to the migration of water molecules from the shale formation into the drilling fluid, which can cause the shale to shrink.

Moreover, when the drilling fluid activity is close to zero, the swelling ratio is defined as the negative extreme swelling ratio of shale (b). The values of b for the different drilling fluids have the following order: $b_{KCl} < b_{CTAB} < b_{SDBS}$. For the KCl drilling fluid, on the one hand, due to the difference in activity between the fluid and the shale, the water molecules in the shale are driven to migrate into the drilling fluid, and consequently the shale gradually contracts; on the other hand, as the radius of the potassium ion is similar to that of the hexagonal cavity in the clay crystal unit, potassium ions can be embedded into the cavity, which can prevent water molecules from entering the shale formation. Therefore, the shale shows the lowest negative extreme swelling ratio with the KCl drilling fluid. The principle of potassium ion embedding to prevent shale swelling is illustrated in Figure 8.

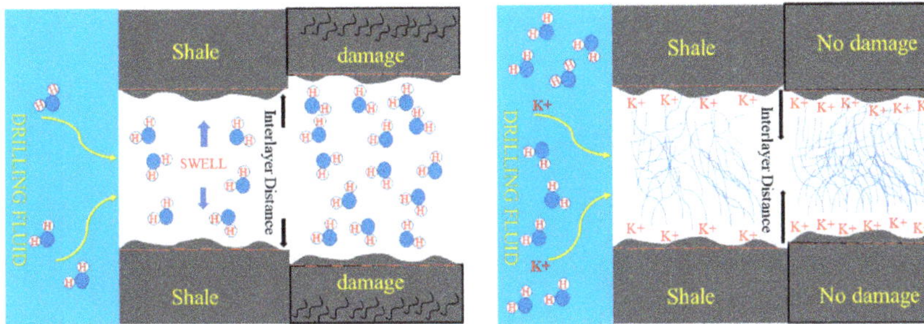

Figure 8. Principle of shale shrinkage due to potassium ion embedding.

3.4. Classification of Safety Levels of Drilling Fluid Activity

The migration of water molecules and ions into a shale formation is greatly affected by the difference between the drilling fluid activity and the shale formation activity (α_s). Therefore, the interaction between the shale formation and the drilling fluid is divided into the following three types [19,29,30]: dehydration action, surface hydration action, and osmotic hydration action. When the drilling fluid activity is higher than the drilling fluid critical activity (α_{wt}), a large amount of water molecules from the drilling fluid migrate into the shale reservoirs, which causes the shale to swell. However, when the drilling fluid activity is slightly lower than the shale formation activity, the resulting slight shrinkage of the shale formation is beneficial to the wellbore stability. Furthermore, when the drilling fluid activity is greatly reduced, the wellbore stability decreases since large amounts of water molecules migrate from the shale into the drilling fluid, causing the activity to reach the dehydration critical activity (α_{wr}). The dehydration critical activity and shale formation activity are defined as the lower limit and upper limit of the safe activity, respectively, and are considered as characteristic parameters that are closely related to the wellbore stability. The range of activities between these two parameters is defined as the safe activity window. Figure 9 is a schematic diagram indicating the divisions of the safety levels of drilling fluid activity.

As shown in Figure 9, the drilling fluid activity and swelling ratio are divided into four zones: the complete dehydration zone (A), the weak dehydration zone (B), the surface hydration zone (C), and the osmotic hydration zone (D). The zones are described as follows:

(1) In zone (A), the drilling fluid activity is lower than the dehydration critical activity. Therefore, the resulting drastic dehydration caused by the loss of free water molecules from the shale formation can result in the serious shrinkage of the shale. As can be seen from Figure 9, the degree of shale shrinkage is largest for the CTAB drilling fluid, followed by the SDBS and the KCl drilling fluids;

(2) In zone (B), the drilling fluid activity is between the dehydration critical activity and the shale formation activity. The dehydration action is weak and the degree of shale shrinkage is minor compared to zone A;

(3) In zone (C), the drilling fluid activity is between the shale formation activity and the critical drilling fluid activity. Due to the fact that water molecules are adsorbed on the surface of the clay through the van der Waals force, shale hydration only occurs on the shale surface, so the swelling ratio of shale is small. As can be seen from Figure 9, the activity window of the surface hydration of the KCl drilling fluid is the largest, followed by the CTAB and SDBS drilling fluids. This means that the surface hydration activity threshold of the KCl drilling fluid is the lowest and that of the SDBS drilling fluid is the highest;

(4) In zone (D), the drilling fluid activity is greater than the drilling fluid critical activity. Therefore, a large amount of water molecules in the drilling fluid are flushed into the shale formation, which causes the clay minerals to swell severely. As can be seen from Figure 9, in this zone, the shale swelling ratio is the largest for the CTAB drilling fluid, which indicates that the swelling-inhibition effect of the drilling fluid on the shale is the worst in this zone.

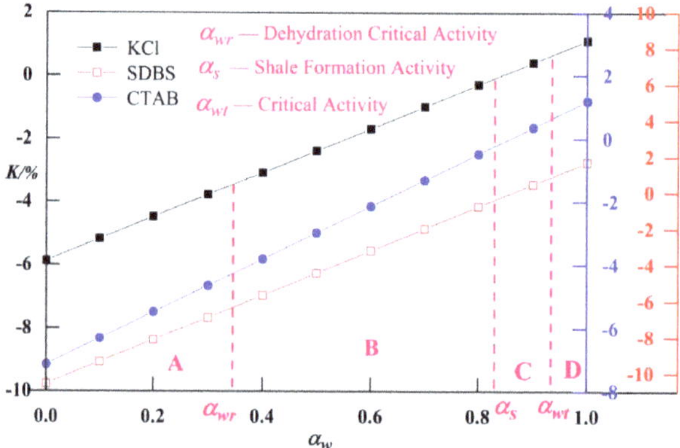

Figure 9. The divisions of the safety levels of drilling fluid activity.

4. Conclusions

In this paper, wettability and swelling tests were conducted to investigate the effects of the activities of three types of drilling fluid on the surface and osmotic hydration of shale. The following conclusions can be drawn:

(1) As the concentration of inhibitors in the drilling fluid system increases, the activity of the CTAB drilling fluid experiences the largest increase, while the activity of the KCl drilling fluid experiences the lowest increase;

(2) As drilling fluid activity increases, a decreasing trend is observed in the rate of change of the contact angles between the fluid and shale, and the shale swelling ratio increases obviously. The KCl drilling fluid showed the best inhibitory effect on shale surface hydration, and also caused the lowest negative extreme shale swelling ratio;

(3) Based on the safety levels of drilling fluid activity, when the KCl drilling fluid was used, the shale underwent the least shrinkage in the complete dehydration zone and the surface hydration activity threshold was the lowest in the surface hydration zone. The largest osmotic hydration zone was observed with the CTAB drilling fluid, which is disadvantageous for wellbore stability.

Author Contributions: H.C. proposed this topic and supervised the work; Z.Z. mainly carried out and completed this work; T.B. provided and analyzed the field data; P.S. and others supplemented and perfected this work in the later stage; all authors read and approved the final manuscript.

Acknowledgments: The work presented in this paper was financially supported by the National Natural Science Foundation of China (Grant No.41302124), Open Funding by Key Laboratory of Oil and Gas Reservoir Geology and Exploitation (Grant No. PLN201607) and Open Funding by Key Laboratory of Drilling and Exploitation Technology in Complex Conditions (Grant No. EDLF2017). We are also grateful to all the editors and reviewers for their invaluable comments.

Conflicts of Interest: The authors declare no conflict of interest.

Abbreviations

KCl	potassium chloride
SDBS	sodium dodecylbenzene sulfonate
CTAB	cetyltrimethylammonium bromide
b	negative extreme swelling ratio
α_w	drilling fluid activity
$\Delta\beta/t$	change ratio of the fluid-shale contact angle
K	shale swelling ratio
a	the ratio of the shale swelling ratio to the drilling fluid activity
α_{wo}	critical drilling fluid activity
α_s	shale formation activity
α_{wt}	drilling fluid critical activity
α_{wr}	dehydration critical activity

References

1. Moslemizadeh, A.; Saeed Khezerloo-ye, A.; Shahbazi, K.; Zendehboudi, S. A triterpenoid saponin as an environmental friendly and biodegradable clay swelling inhibitor. *J. Mol. Liq.* **2017**, *247*, 269–280. [CrossRef]
2. Zeynali, M.E. Mechanical and physical-chemical aspects of wellbore stability during drilling operations. *J. Pet. Sci. Eng.* **2012**, *82*, 120–124. [CrossRef]
3. Shadizadeh, S.R.; Moslemizadeh, A.; Dezaki, A.S. A novel nonionic surfactant for inhibiting shale hydration. *Appl. Clay Sci.* **2015**, *118*, 74–86. [CrossRef]
4. Steiger, R.P.; Leung, P.K. Quantitative Determination of the Mechanical Properties of Shales. *SPE Drill. Eng.* **1992**, *7*, 181–185. [CrossRef]
5. Ye, Y.; Chen, S.; Wang, Z.; Yang, X.; Peng, Y.; Cai, J.; Nasr-El-Din, H.A. Improving wellbore stability of shale by adjusting its wettability. *J. Pet. Sci. Eng.* **2017**, *161*, 692–702. [CrossRef]
6. Diaz-Perez, A.; Cortes-Monroy, I.; Roegiers, J. The role of water/clay interaction in the shale characterization. *J. Pet. Sci. Eng.* **2007**, *58*, 83–98. [CrossRef]
7. Luo, Z.; Wang, L.; Yu, P.; Chen, Z. Experimental study on the application of an ionic liquid as a shale inhibitor and inhibitive mechanism. *Appl. Clay Sci.* **2017**, *150*, 267–274. [CrossRef]
8. Liang, L.; Xiong, J.; Liu, X. Effects of hydration swelling and wettability on propagation mechanism of shale formation crack. *Pet. Geol. Exp.* **2014**, *36*, 780–786.
9. Wang, G. Hydration Characteristics of Hard Brittle Shale and Its Influence on Collapse Pressure. Master Thesis, Southwest Petroleum University, Chengdu, China, 2014. Available online: http://cdmd.cnki.com.cn/Article/CDMD-10615-1018021314.htm (accessed on 15 August 2019).
10. Chenevert, E.; Shale, M. Alteration by Water Adsorption. *J. Pet. Technol.* **1970**, *22*, 1141–1148. [CrossRef]
11. Talal, M.A.; Kuwait, U.; Zhang, J.; Baker, A.; Martin, E.C.; Mukul, M. Factors Controlling the Membrane Efficiency of Shales When Interacting with Water-Based and Oil-Based Muds. In Proceedings of the SPE International Oil & Gas Conference and Exhibition, Beijing, China, 5–7 December 2006; Available online: https://www.onepetro.org/conference-paper/SPE-100735-MS (accessed on 15 August 2019).
12. Van Oort, E. On the physical and chemical stability of shales. *J. Pet. Sci. Eng.* **2003**, *38*, 213–235. [CrossRef]
13. Tang, W. Study on Effects of Shale Hydration on Wellbore Stability. Ph.D. Thesis, China University of Petroleum (East China), Qingdao, China, 2011. Available online: http://cdmd.cnki.com.cn/Article/CDMD-10425-1011287063.htm (accessed on 15 August 2019).
14. Ding, R.; Li, J. The Effect of Water Activity and Semipermeable Membrane on Shale Hydration. *Drill. Fluid Complet. Fluid* **1994**, 23–28. Available online: http://www.cnki.com.cn/Article/CJFDTotal-ZJYW199403004.htm (accessed on 15 August 2019).

15. Al-Bazali, T.; Rakf, M.A. The Role of Chemical Potential and Molecular Diffusion on the geo-Mechanical Stability of Shale. In Proceedings of the 2018 World Congress on Advances in Civil, Environmental, & Materials Research (ACEM18), Incheon, Korea, 27 31 August 2018; Available online: http://www.i-asem.org/publication_conf/acem18/2.ICGE18/W3B.4.GE1160_5034F1.pdf (accessed on 15 August 2019).
16. Chenevert, E.; Shale, M. Control with Balanced-Activity Oil-Continuous Muds. *J. Pet. Technol.* **1970**, *22*, 1309–1316. [CrossRef]
17. Wen, H.; Chen, M.; Jin, Y.; Wen, Z.; Ye, X.; Yang, S. Experimental research on brittle shale failure caused by drilling fluid activity. *Oil Drill. Prod. Technol.* **2014**, 57–60. [CrossRef]
18. Yew, H.C.; Wang, L.C.; Chenevert, E.M.A. Theory on Water Activity Between Drill-Fluid and Shale. In Proceedings of the Rock Mechanics, Tilerson & Wawersik, Santa Fe, NM, USA, 3–5 June 1992. Available online: https://www.onepetro.org/conference-paper/ARMA-92-0717 (accessed on 15 August 2019).
19. Huang, W.; Li, X.; Qiu, Z.; Jia, J.; Wang, Y.; Li, X. Inhibiting the surface hydration of shale formation using preferred surfactant compound of polyamine and twelve alkyl two hydroxyethyl amine oxide for drilling. *J. Pet. Sci. Eng.* **2017**, *159*, 791–798. [CrossRef]
20. Du, D.; Fan, S.; Chenevert, M.E. Study on the osmotic hydration of shale in water based fluid. *Drill. Fluid Complet. Fluid* **1996**, 5–10. Available online: http://www.cnki.com.cn/Article/CJFDTOTAL-ZJYW199603001.htm (accessed on 15 August 2019).
21. Liu, J.; Sun, J. Effects of Drilling Fluid Activity on Hydration and Dispersion of Formation Rocks in Shale Gas Drilling in Chuan-Dian Area. *Drill. Fluid Compl. Fluid* **2016**, *33*, 31–35. [CrossRef]
22. Cai, J.; Yue, Y.; Cao, J.; Yang, X.; Wu, X. Experimental study on the effect of drilling fluid wettability on shale wellbore stability. *J. China Coal Soc.* **2016**, *41*, 228–233. [CrossRef]
23. Moslemizadeh, A.; Shadizadeh, S.R. Minimizing water invasion into kazhdumi shale using nanoparticles. *Iran. J. Oil Gas Sci. Technol.* **2015**, *4*, 15–32. Available online: http://ijogst.put.ac.ir/article_12475_a66037b9fcaa4c236494db533a218905.pdf (accessed on 15 August 2019).
24. Liu, X.; Xiong, J.; Liang, L.; Luo, C.; Zhang, A. Analysis of the wettability of Longmaxi Formation shale in the south region of Sichuan Basin and its influence. *Nat. Gas Geosci.* **2017**, *161*, 692–702. [CrossRef]
25. Shi, X.; Wang, L.; Guo, J.; Su, Q.; Zhuo, X. Wettability, oil recovery, and interfacial tension with an SDBS–dodecane–kaolin system. *J. Colloid Interface Sci.* **1999**, *214*, 368–372. [CrossRef]
26. Shi, X.; Wang, L.; Guo, J.; Su, Q.; Zhuo, X. Effects of inhibitor KCl on shale expansibility and mechanical properties. *Petroleum* **2018**. [CrossRef]
27. Moslemizadeh, A.; Aghdam, S.K.; Shahbazi, K.; Aghdam, H.K.; Alboghobeish, F. Assessment of swelling inhibitive effect of CTAB adsorption on montmorillonite in aqueous phase. *Appl. Clay Sci.* **2016**, *127*, 111–122. [CrossRef]
28. Cao, H.; Wang, T.; Bao, T.; Sun, P.; Zhang, Z.; Wu, J. Effective Exploitation Potential of Shale Gas from Lower Cambrian Niutitang Formation, Northwestern Hunan, China. *Energies* **2018**, *11*. [CrossRef]
29. Liu, X.; Liu, K.; Gou, S.; Liang, L.; Cheng, L.; Guo, Q. Water-Soluble Acrylamide Sulfonate Copolymer for Inhibiting Shale Hydration. *Ind. Eng. Chem. Res.* **2014**, *53*, 2903–2910. [CrossRef]
30. Wang, L.; Liu, S.; Wang, T.; Sun, D. Effect of poly(oxypropylene)diamine adsorption on hydration and dispersion of montmorillonite particles in aqueous solution. *Colloids Surf. Physicochem. Eng. Asp.* **2011**, *381*, 41–47. [CrossRef]

© 2019 by the authors. Licensee MDPI, Basel, Switzerland. This article is an open access article distributed under the terms and conditions of the Creative Commons Attribution (CC BY) license (http://creativecommons.org/licenses/by/4.0/).

Article

Study on the Mechanism of Ionic Stabilizers on Shale Gas Reservoir Mechanics in Northwestern Hunan

Pinghe Sun [1,2], Junyi Zhu [1,2], Binkui Zhao [1,2], Xinxin Zhang [1,2,*], Han Cao [1,2,*], Mingjin Tian [3], Meng Han [1,2] and Weisheng Liu [1,2]

1 Key Laboratory of Metallogenic Prediction of Nonferrous Metals and Geological Environment Monitoring, Ministry of Education, Changsha 410083, China; pinghesun@csu.edu.cn (P.S.); zhujunyi@csu.edu.cn (J.Z.); Beanque515@yahoo.com (B.Z.); hanmeng@csu.edu.cn (M.H.); weishengliu@csu.edu.cn (W.L.)
2 School of Geosciences and Info–physics, Central South University, Changsha 410083, China
3 Hunan Provincial Communications Planning, Survey and Design Institute Co., Ltd., Hunan Changsha 410008, China; mingjin-TIAN@163.com
* Correspondence: zhangxx@csu.edu.cn (X.Z.); hancao@csu.edu.cn (H.C.)

Received: 27 April 2019; Accepted: 24 June 2019; Published: 25 June 2019

Abstract: The shale of the lower Cambrian Niutitang formation in northwestern Hunan is an ideal reservoir for shale gas. There is a close connection between borehole stability and drilling fluid in shale gas drilling. Ionic stabilizer is a new type of stratum consolidation agent that inhibits the hydration expansion of clay minerals and improves mechanical strength of the borehole. The traditional idea of pore wall protection is to use drilling fluid additives to prevent shale from interacting with water. However, ionic stabilizer can change the hydrophilic of clay minerals in shale, making the particles become hydrophobic and dense, therefore, the formation stability can be enhanced simultaneously. The material used in this paper is different from the normal ionic stabilizer, some chemical bonds that have been changed in the new material called enhanced normality ionic (ENI) stabilizer. This paper utilized the shale samples those obtained from Niutitang formation to study the connection between ENI and the mechanical properties of shale. Mechanical tests and microscopic pore tests were performed on different samples which were soaked in water and the ENI with different concentrations. It has been found through tests that ENI can inhibit the development of shale pores, and as the concentration increases, the inhibition increases. In addition, as the ENI concentration increases, the uniaxial compressive strength and Young's modulus of the shale increase, and the ratio of stability coefficients decreases. It can be concluded that the ENI can improve the mechanical strength of carbon shale, and prevent the development of rock damage. Moreover, it can improve the ability of rock to resist damage, and enhance borehole stability initiatively.

Keywords: ionic stabilizer; borehole stability; shale drilling fluid; shale gas reservoir; Niutitang formation

---

1. Introduction

Shale gas is an unconventional energy source with great potential for exploration and development. The black carbonaceous shale of the lower Cambrian Niutitang formation is one of the key main targets for the exploration and development of shale gas in the future in China [1–3]. It is widely distributed in northwestern Hunan. The shale in this area has thick sediment and high abundance of organic matter, which has abundant shale gas resources [4,5]. The interaction between drilling fluid and shale leads to wellbore instability, which has been a concern in shale gas development [6–9]. Furthermore, the productivity of the wells decreases due to this instability, which also increases the drilling cost [10]. In order to improve the stability of the borehole, scholars added some drilling additives to inhibit shale hydration. For example, some scholars used polymer to reduce water loss and inhibited its hydration expansion in shale drilling [11–14]. Some scholars use nano-compounds to maintain drilling stability to seal small holes in shale [15–17]. Other scholars utilized shale inhibitors such as surfactant, polyether and ammonium salt to improve the stability of the pore walls [18–20]. These additives improved the stability of the pore walls by preventing the shale from interacting with water. However, these are some passive defense modes. In fact, there is a better way to enhance shale strength and improve borehole wall stability actively, that is, by using ionic stabilizers.

Ionic stabilizer is one kind of environmental engineering materials, which is composed of inorganic and organic materials [21]. Ionic stabilizer can change the hydrophilic of cations in clay minerals [22]. In addition, it can reduce the thickness of bound water film and enhance the stratum overall strength [23–26]. It emerged in the 1970s and has been widely used in various industries, and currently in civil engineering. However, the application of ionic stabilizer in shale drilling is still rare. There is a close connection between borehole stability and drilling fluid in shale gas drilling [27–29]. The traditional borehole stability protection idea is to reduce the drilling fluid pressure and stratum intrusion [12,30]. However, ionic stabilizer can change the hydrophilic of clay minerals in shale, making the particles become hydrophobic and dense, therefore, the formation stability can be enhanced. The material used in this paper is different from the normal ionic stabilizer, some chemical bonds that have been changed in the new material called enhanced normality ionic (ENI) stabilizer. This paper utilized the shale that obtained from the Niutitang formation to explore the connection between ENI and the mechanical properties of shale.

Mechanical tests and microscopic pore tests were selected as two different mechanical methods in this study [31–33]. Uniaxial compression test and pore characteristics observation were performed on different samples which were soaked in water and the ENI with different concentrations. The mechanism of ENI on the carbonaceous shale samples were analyzed based on the test results.

2. Materials and Methods

2.1. Materials

ENI used in this study was light yellow, mainly composed of a petroleum sulfonated agent, modified sodium silicate, and a modified polymer surface active agent. It can decompose the clay mineral and nonclay particles in the stratum. Moreover, it can recrystallize the clay mineral particles and reduce the invasion of filtrate relying on making use of bonding and winding between polymer compounds and leaving the particles hydrophobic and dense [34,35].

The black carbonaceous shale samples utilized in this paper were obtained from the lower Cambrian Niutitang formation in northwestern Hunan. Most shale gas can be explored in this stratum. Reservoir samples with diameter 50 mm (Φ50) and length 100 mm were obtained by the small modified portable drilling rig in the field (Figure 1). Other samples were cut into 15 × 10 × 5 mm in the lab. However, in order to eliminate the influence of man-made disturbance on rock samples, the intact samples were selected to perform in all experiments. Mineral composition of the shale was obtained by X–ray diffraction (XRD) analysis of rock sample fragments, and the results are indicated in Table 1. The results showed that clay minerals were mainly strong-expanded smectite. There was more than 25% smectite in the samples which indicates that these rocks were prone to changes in strength due to the hydration expansion.

Figure 1. Small modified portable drilling rig in the jobsite.

Table 1. X-ray diffraction (XRD) analysis results of mineral composition of carbonaceous shale.

Rock	Mass Percentage					
	Quartz	Mica	Feldspar	Smectite	Anatase	Amorphous Substance
Carbonaceous shale	16.57	13.64	10.82	25.58	5.26	28.13

2.2. Uniaxial Compression Test

Uniaxial compression tests were performed after immersing the samples in different concentrations of ENI. The ENI concentrations in the experiment were 0%, 1%, 2%, 3%, and 4%. The uniaxial compression deformation tests were carried out by a rock tensile splitting machine, and the dynamic static strain gauge was used for deformation monitoring at the same time. The shale samples in the experiment were immersed in different concentrations of ENI for 24 hours at constant temperature (25 ± 2 °C). Continuous loading was added by the rate of 0.5MPa/s. The data was recorded automatically until the samples were destroyed. The stress-strain curve was measured by the machine, and the uniaxial compressive strength (UCS) and Young's modulus of the shale were calculated based on the curve.

2.3. Microscopic Pore Test

Microscopic pore tests were performed after immersing the samples in different concentrations of ENI. The ENI concentrations in the test were 0%, 1%, 2%, 3%, and 4%. Pore characteristics were analyzed by scanning electron microscope (SEM). SEM image information was processed by Photoshop and ImageJ2x image processing software. The ratio of fractal dimension and stability coefficient of shale were obtained by analysis and fitting.

2.3.1. Fractal Dimension

Fractal dimension is a measure of the irregularity of complex shapes. It can identify the quantitative description of the complexity and heterogeneity on pore structure [36]. The larger the fractal dimension, the higher the spatial geometric complexity of the pore shape and the rougher the sample surface. Therefore, the stability of shale can be evaluated by fractal dimension. Basic shape characteristic parameters such as area, perimeter of the shale sample can be obtained by processing the SEM image. Based on the definition of damage mechanics, fractal dimension can be calculated by perimeter and pore area as follows:

$$\lg(C) = \frac{I}{2} \times \lg(A) + b, \tag{1}$$

where C is the pore perimeter, the unit is mm, A is the pore area, the unit is mm^2, I is the pore system fractal dimension, and b is a constant.

2.3.2. Ratio of Stability Coefficients

The relative damage variable can be defined as damage value according to the damage theory. Damage value is a "deterioration coefficient" that affects the rock mechanical properties seriously [37–39]. Pores and fractures in the shale reservoir will weaken the overall strength and reduce the stability. The smaller damage value is, the better physical and mechanical properties of rock are, and the better borehole stability is. Damage value can be calculated by the total micropores area and the bearing area of the shale sample as follows:

$$D = \frac{S_i}{S_0}, \tag{2}$$

where: D is the damage value; S_i is the sum of the area occupied by the micropores on the bearing surface of the rock sample, the unit is mm^2; S_0—the apparent bearing area of the rock sample, the unit is mm^2.

The stability coefficient is an important indicator for evaluating borehole stability and rock strength. The smaller stability coefficient is, the better physical and mechanical properties of rock are, and the more borehole stability is. According to the deification of damage tolerance, the stability coefficient "F" can be calculated as follows:

$$D = \frac{S_i}{S_0}, \tag{3}$$

The ratio of rock stability coefficients can be derived based on Equations (2) and (3), which is the ratio under the effect of different concentrations of ENI. The ratio of the stability coefficient K_n is calculated as follows:

$$K_n = \frac{S_{w_n}}{S_{w_0}}, \tag{4}$$

where S_{w_0} is the sum of the area occupied by micro–pores on the surface of the sample soaked in water, the unit is μm^2; S_{w_n} is the sum of the area occupied by micro–pores on the surface of the rock soaked by n%ENI, the unit is μm^2; K_n is the ratio of water to n%ENI stability coefficient.

In order to determine the strength and stability of the shale sample under the influence of water and different concentrations of ENI, the ratio of the stability coefficient is an intuitive expression that can reflect pore, strength and stability properties. The smaller the value, the larger the difference between the two shale samples and the higher the intensity of the shale sample.

3. Results and Discussions

3.1. Uniaxial Compression Test

The relationship curves between axial compression stress–strain and radial strain are indicated in Figure 2. It was derived from the experimental data.

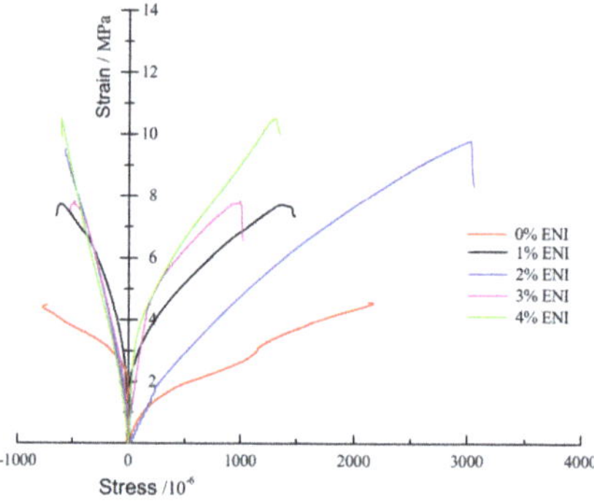

Figure 2. Uniaxial compression stress–strain curve of carbonaceous shale.

The UCS and Young's modulus values (shown in Figure 3) of samples under the different concentrations of ENI were obtained through analysis and calculation. These results were based on the linear part of curve in Figure 2.

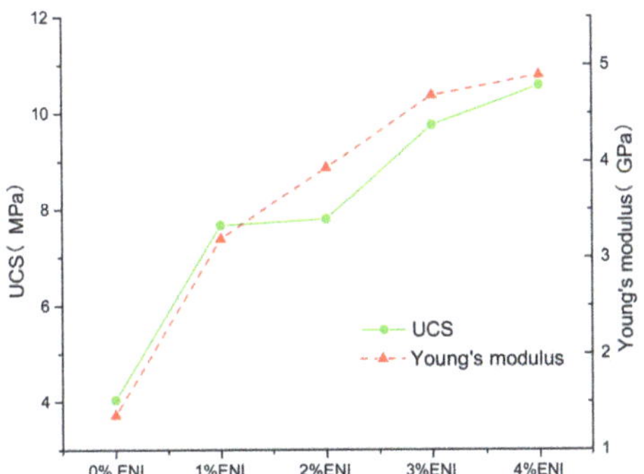

Figure 3. Shale uniaxial compression test results.

The results indicate that UCS was positively correlated with the concentration of ENI. When the ENI concentration was 0%, UCS reached the minimum, equal to 4.045MPa. However, when the ENI concentration was 4%, the UCS reached the maximum, equal to 10.583MPa, which was 2.61 times the minimum value. Simultaneously, the Young's modulus was also positively correlated with ENI concentration. When the ENI concentration was 0%, Young's modulus reached the minimum, equal to 1.361GPa. When the ENI concentration was 4%, its value reached the maximum, equal to 4.897GPa, which was 3.59 times the minimum value. The higher the ENI concentration is, the better the mechanical properties of the shale sample are. In addition, for these two parameters, the growth rate of Young's modulus was greater than UCS when the ENI concentration increments were the same. This indicates that the inhibitory effect of ENI on shale hydration expansion is more effective than the increase of overall strength. This shows that ENI can enhance the borehole stability of shale gas drilling. As concentration of ENI increases, the curve gradually slows down, but no vertices appear. The curve indicates that as ENI concentration increases, UCS and Young's modulus increase rate of the shale decreases. As one of the concerns, ENI suitable concentration problem needs to be solved in a future study.

3.2. *Pore Characteristics*

Other square samples (length 15 × wide 10 × height 5 mm) were immersed in different concentrations of ENI for 24 hours. Then the surface morphology of samples was analyzed by SEM images. SEM images are first chosen to address scaling issues when choosing samples for 2D and 3D volumetric analysis. The images' microstructural information is indicated in Figure 4.

In order to obtain the quantitative information of shale pores and fissures, 500x magnification SEM images were processed by Photoshop and ImageJ2x. After smoothing, the resultant images, which appeared similar to the original photomicrograph but with a minimal background, were then converted to binary images by setting a threshold. In order to unify data, pore characteristics data of the rock samples were summarized. The average values of the pore area and perimeter of the shale sample under different concentrations of ENI were calculated from the summarized data (shown in Table 2).

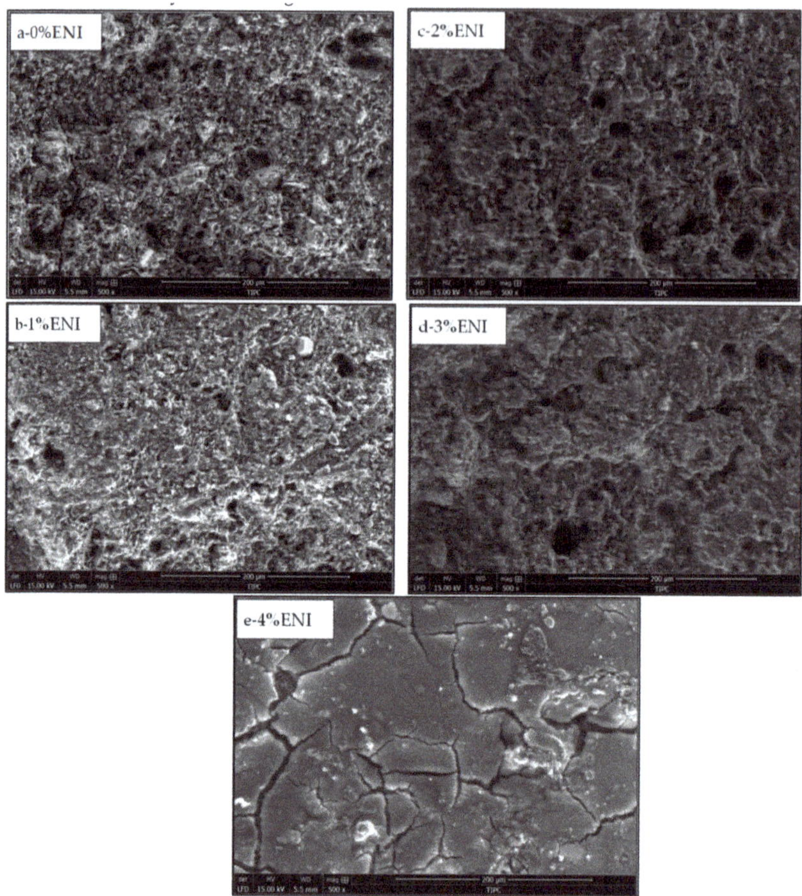

Figure 4. Microstructural diagram of shale pores. (**a**) 1%ENI; (**b**) 1% ENI; (**c**) 2% ENI; (**d**) 3% ENI; (**e**) 4% ENI.

Table 2. Statistical table of information on shale pore characteristics.

	Pore Area/μm²					Pore Perimeter/μm				
	0% ENI	1% ENI	2% ENI	3% ENI	4% ENI	0% ENI	1% ENI	2% ENI	3% ENI	4% ENI
Summary	84.716	50.656	47.454	38.875	36.278	1955.551	589.734	656.714	204.437	166.272
Average	0.589	0.418	0.418	0.368	0.359	5.477	2.657	2.433	1.573	1.543

The pore variation on shale sample surface under different conditions is clear when the qualitative observation of shale microscopic morphology is obtained in Figure 3. When water was immersed into the sample, its structure became very loose due to the extension of pores and cracks. Otherwise, with the increasing of ENI concentration, the pores and fractures in the same samples decrease and the rock surface becomes smooth gradually. For example, only a few pores can be seen on the surface of the sample after 4% ENI added from Figure 4e. At the same time, according to Table 2, as the ENI concentration increases, the pore area and perimeter in the sample decrease. This trend was similar to the SEM results, indicating that ENI can discourage the development of shale pore. In order to further study the relationship between ENI and the mechanical properties of shale, the ratio of fractal dimension to stability coefficient is obtained from the pore characteristic data. The ratio of the stability coefficients derived from the stability coefficients is affected by different concentration water and ENI.

3.2.1. Fractal Dimension

In this paper, the experimental results of the microscopic pore characteristics in the samples are fitted by Equation (1) (shown in Figure 5). In the figure, the logarithm of area was set as the abscissa and the logarithm of the perimeter is set as the ordinate. If the experimental results were linearly correlated, the fitting results were good. Then, the value of fractal dimension is equal to twice the line slope.

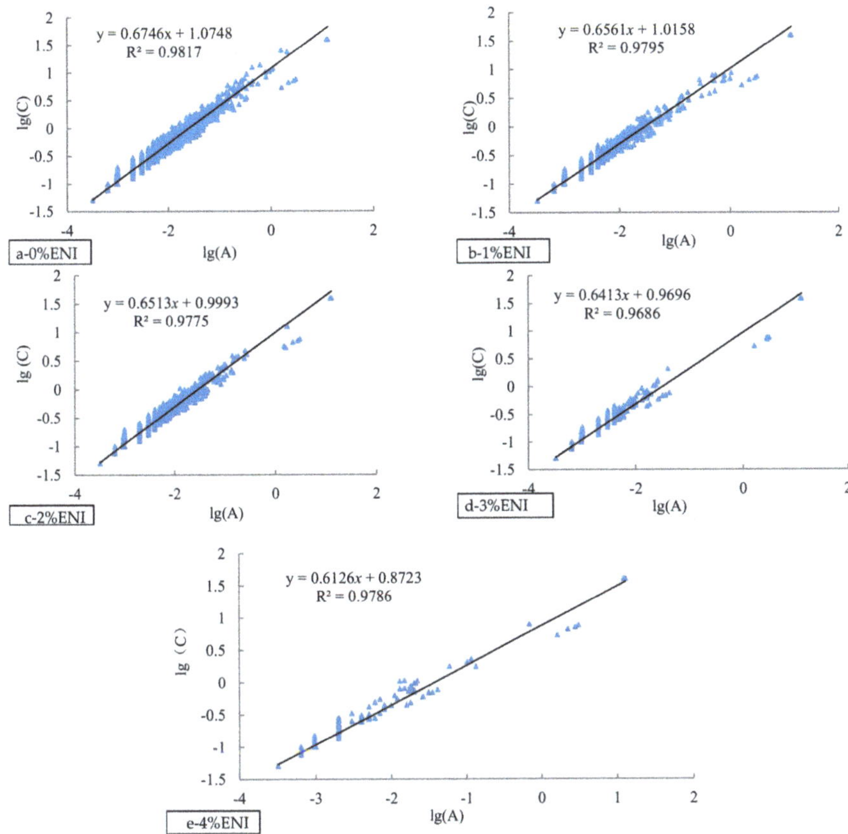

Figure 5. Shale sample fractal curve fitting result chart. (**a**) 1% ENI; (**b**) 1% ENI; (**c**) 2% ENI; (**d**) 3% ENI; (**e**) 4% ENI.

It can be observed from Figure 5 that lg (A) and lg (C) were positively correlated, and the linear correlation between parameters is good. The judgment coefficient R^2 belongs to 0.9686~0.9817, so the fitting goodness is very high. Therefore, the fractal dimension can be derived from the calculation of the fitted curve. The fractal dimension of the sample could be obtained in Table 3.

Table 3. Fractal dimension with different enhanced normality ionic (ENI) concentrations.

Concentration	Fractal Dimension
0% ENI	1.35
1% ENI	1.31
2% ENI	1.30
3% ENI	1.28
4% ENI	1.23

Fractal dimension can represent the spatial geometric complexity of pore shape. The larger the fractal dimension, the higher the spatial geometric complexity of the pore shape, the rougher the sample surface, and the worse the mechanical properties of the sample [40]. There is a certain correlation between UCS and pores. In the same rock sample, the more the rock pores, the smaller the UCS. The effect of the pores on the uniaxial compressive strength is greater than the Young's modulus [41,42]. According to Table 3, when the ENI concentration was 0%, the fractal dimension was 1.35. While the concentration increases by 1%, the fractal dimension decreases to 1.31, which is 3% lower than the former one. ENI can inhibit shale pore development. However, the inhibitory effect was not significant when the ENI concentration was 1%. With the increase of ENI concentration, the fractal dimension decreases gradually. It indicates that drilling fluid inhibition increases gradually. Meanwhile, other parameters like pore extension and rock damage are inhibited, and the sample surface becomes smooth. The most important is that the shale sample strength increases. This fractal dimension is opposite to the trend of UCS, indicating that ENI enhances shale strength by inhibiting pore growth. It also shows that the shale strength is negatively correlated with the pores. Under the influence of 4% of ENI, the fractal dimension reached the minimum value of 1.23, which is 9.19% lower than 0% of ENI. It indicates that 4% of ENI had the strongest inhibitory effect on shale. When the ENI was in contact with the shale, it changed the hydrophilicity of the shale surface and made the shale pores smaller. At the same time, ENI made the shale particles dense and the sample strength will increase. As the increases of ENI concentration, the rock integrity was better. More importantly, the mechanical properties of the shale were improved.

3.2.2. Ratio of Stability Coefficients

The ratio of stability coefficients K_n can be obtained by Equation 4 at different concentrations of ENI. The statistical range was calculated according to the 50 mm diameter core samples. The K_n values ($n = 1, 2, 3, 4$) are shown in Table 4.

Table 4. Ratio of shale stability.

Number	K_1	K_2	K_3	K_4
Ratio of Shale Stability	0.598	0.560	0.459	0.428

K_n represents the difference in shale strength between ENI and without ENI. Under the action of water, the shale strength decreases. ENI can change the hydrophilicity of shale, reduce the impact of water on shale and increase sample's strength. The larger K_n, the different concentrations' ENI affection on shale strength is greater. It also means that the shale strength is higher under the action of ENI. According to Table 4, for the four different ENI concentrations, the ratio of stability coefficients was less than 0.6, much less than 1. The stability of the sample under ENI was higher than that of the sample without ENI, and the shale pore extension was inhibited. It indicates that ENI can not only improve the stability of the borehole wall, but also enhance the strength of shale. The ratio of stability coefficient was negatively correlated with ENI concentration. It indicates that the higher the ENI concentration, the better the sample integrity. Meanwhile, the sample strength was also higher and the pore walls are more stable. K_4 was only 0.428, which was 28% smaller than K_1, and K_1 was 0.598. ENI of 4% had a better effect on shale strength than 1% ENI. These results indicate that the shale sample surface is further inhibited and its strength increases as ENI concentration increases. In addition, the difference between K_3 and K_2 was the largest, which indicates that the shale strength changes most when the ENI concentration was 3%. The reason for this change cannot be explained detail based on available data, which may require more trials to verify.

By comparison, it was found that the results of the uniaxial compression test and the pore microscopic characteristics test were consistent. As the concentration of ENI increases, the pores decrease and the surface of the shale becomes smooth. It shows that ENI enhances shale strength and improves pore wall stability by inhibiting shale pore development. It was further verified that ENI could improve the mechanical strength of shale and effectively improve the stability of the borehole.

4. Conclusions

(1) The UCS and Young's modulus are positively correlated with the ENI concentration. As the ENI concentration increases, shale hydration expansion is inhibited and the overall strength of the shale increases. Due to the inhibitory effect on clay minerals with ENI, the effect of ENI on Young's modulus is greater than that of UCS. As the concentration of ENI increases, the UCS and Young's modulus increase rates of shale decrease, but no apex appears. One issue is to determine the appropriate concentration of ENI in future studies.

(2) When the ENI is in contact with the shale, it changes the hydrophilicity of the shale surface and makes the shale pores smaller. At the same time, ENI will make the shale particles become dense and the sample strength will increase. With the increase of ENI concentration, the fractal dimension decreases gradually. It indicates that drilling fluid inhibition increases gradually. Meanwhile, other parameters like pore extension and rock damage are inhibited, and the sample surface becomes smooth. What matters most is that the shale sample strength increases. This fractal dimension is opposite to the trend of UCS, indicating that ENI enhances shale strength by inhibiting pore growth.

(3) The ratio of stability coefficient is an intuitive expression which can reflect pore, strength and stability properties. The ratio of the stability coefficient is inversely related to the ENI concentration. This means that the higher the ENI concentration, the better the sample integrity. At the same time, the sample strength is also higher and the borehole is more stable. The stability of the sample under ENI is higher than the sample without ENI, and the shale pore extension is inhibited. It indicates that ENI can not only improve the stability of the borehole wall, but also enhance the strength of shale.

Author Contributions: Conceptualization, H.P., J.Z., and M.T.; Methodology, H.P., H.C. and X.Z.; Software, M.T. and J.Z.; Validation, B.Z., W.L., J.Z. and M.H.; Writing-Original Draft Preparation, H.P., J.Z., X.Z. and H.C.; Writing-Review & Editing, H.P., J.Z., X.Z., B.Z., W.L., and H.C.

Funding: This study was supported by the National Natural Science Foundation of China (No. 41602372), open fund of the State Key Laboratory of Oil and Gas Reservoir Geology and Exploitation (Southwest Petroleum University, No. PLN201609 and No. PLN201607, fund of Ministry of Land and Resources Key Laboratory of Drilling Technology in Complex Conditions (No. DET201612), and the Fundamental Research Funds for the Central Universities of Central South University (No. 2018zzts690 and No. 2019zzts636). All of the authors would like to direct their thanks to the professors and students in the Key Laboratory of Metallogenic Prediction of Nonferrous Metals and Geological Environment Monitoring (Central South University), Ministry of Education.

Conflicts of Interest: The authors declare no conflict of interest.

References

1. Tonglou, G. Key geological issues and main controls on accumulation and enrichment of Chinese shale gas. *Pet. Explor. Dev.* **2016**, *43*, 349–359.
2. Wan, Y.; Tang, S.; Pan, Z. Evaluation of the shale gas potential of the lower Silurian Longmaxi Formation in northwest Hunan Province, China. *Mar. Pet. Geol.* **2017**, *79*, 159–175. [CrossRef]
3. Ge, H.-K.; Yang, L.; Shen, Y.-H.; Ren, K.; Meng, F.-B.; Ji, W.-M.; Wu, S. Experimental investigation of shale imbibition capacity and the factors influencing loss of hydraulic fracturing fluids. *Pet. Sci.* **2015**, *12*, 636–650. [CrossRef]

4. Cao, H.; Wang, T.; Bao, T.; Sun, P.; Zhang, Z.; Wu, J. Effective exploitation potential of shale gas from lower cambrian niutitang formation, Northwestern Hunan, China. *Energies* **2018**, *11*, 3373. [CrossRef]
5. Chen, J.; Cao, H.; Sun, P. Fracability evaluation of shale in Niutitang formation in northwestern Hunan. *Earth Sci. Front.* **2017**, *24*, 390–398. (in Chinese).
6. Kumar, S.; Jain, R.; Chaudhary, P.; Mahto, V. Development of inhibitive water based drilling fluid system with synthesized graft copolymer for reactive Indian shale formation. In *SPE Oil & Gas India Conf.Exhibit*; Society of Petroleum Engineers: Dubai, UAE, 2017.
7. Liu, X.; Zeng, W.; Liang, L.; Lei, M. Wellbore stability analysis for horizontal wells in shale formations. *J. Nat. Gas Sci. Eng.* **2016**, *31*, 1–8. [CrossRef]
8. Abbas, A.K.; Flori, R.E.; Al-Anssari, A.; Alsaba, M. Laboratory analysis to assess shale stability for the Zubair Formation, Southern Iraq. *J. Nat. Gas Sci. Eng.* **2018**, *56*, 315–323. [CrossRef]
9. Yu, M.; Chenevert, M.E.; Sharma, M.M. Chemical–mechanical wellbore instability model for shales: Accounting for solute diffusion. *J. Pet. Sci. Eng.* **2003**, *38*, 131–143. [CrossRef]
10. Mahto, V.; Sharma, V.P. Rheological study of a water based oil well drilling fluid. *J. Pet. Sci. Eng.* **2004**, *45*, 123–128. [CrossRef]
11. Xuan, Y.; Jiang, G.; Li, Y.; Geng, H.; Wang, J. A biomimetic drilling fluid for wellbore strengthening. *Pet. Explor. Dev.* **2013**, *40*, 531–536. [CrossRef]
12. Ferreira, C.C.; Teixeira, G.T.; Lachter, E.R.; Nascimento, R.S.V. Partially hydrophobized hyperbranched polyglycerols as non–ionic reactive shale inhibitors for water–based drilling fluids. *Appl. Clay Sci.* **2016**, *132*, 122–132. [CrossRef]
13. Xu, J.-G.; Qiu, Z.; Zhao, X.; Huang, W. Hydrophobic modified polymer based silica nanocomposite for improving shale stability in water–based drilling fluids. *J. Pet. Sci. Eng.* **2017**, *153*, 325–330.
14. Koteeswaran, S.; Pashin, J.C.; Ramsey, J.D.; Clark, P.E. Quantitative characterization of polyacrylamide-shale interaction under various saline conditions. *Pet. Sci.* **2017**, *14*, 586–596. [CrossRef]
15. Zhang, J.; Li, L.; Wang, S.; Wang, J.; Yang, H.; Zhao, Z.; Zhu, J.; Zhang, Z. Novel micro and nano particle-based drilling fluids: Pioneering approach to overcome the borehole instability problem in shale formations. In *SPE Asia Pacific Unconventional Resources Conference and Exhibition*; Society of Petroleum Engineers: Dubai, UAE, 2015.
16. Zulfiqar, S.; Sarwar, M.I.; Rasheed, N.; Yavuz, C.T. Influence of interlayer functionalization of kaolinite on property profile of copolymer nanocomposites. *Appl. Clay Sci.* **2015**, *112*, 25–31. [CrossRef]
17. Vryzas, Z.; Kelessidis, V.C. Nano-based drilling fluids: A review. *Energies* **2017**, *10*, 540. [CrossRef]
18. Zhong, H.; Qiu, Z.; Huang, W.; Sun, D.; Zhang, D.; Cao, J. Synergistic stabilization of shale by a mixture of polyamidoamine dendrimers modified bentonite with various generations in water-based drilling fluid. *Appl. Clay Sci.* **2015**, *114*, 359–369. [CrossRef]
19. Moslemizadeh, A.; Aghdam, S.K.-Y.; Shahbazi, K.; Zendehboudi, S. A triterpenoid saponin as an environmental friendly and biodegradable clay swelling inhibitor. *J. Mol. Liq.* **2017**, *247*, 269–280. [CrossRef]
20. Jain, R.; Mahto, V. Formulation of a water based drilling fluid system with synthesized graft copolymer for troublesome shale formations. *J. Nat. Gas Sci. Eng.* **2017**, *38*, 171–181. [CrossRef]
21. Katz, L.E.; Rauch, A.F.; Liljestrand, H.M.; Harmon, J.S.; Shaw, K.S.; Albers, H. Mechanisms of soil stabilization with liquid ionic stabilizer. *Transp. Res. Record* **2001**, *1757*, 50–57. [CrossRef]
22. Berry, S.L.; Boles, J.L.; Brannon, H.D.; Beall, B.B. Performance evaluation of ionic liquids as a clay stabilizer and shale inhibitor. In *SPE International Symposium and Exhibition on Formation Damage Control*; Society of Petroleum Engineers: Dubai, UAE, 2008.
23. Petry, T.M.; Das, B. Evaluation of chemical modifiers and stabilizers for chemically active soils—Clays. *Transp. Res. Rec. J. Transp. Res. Board* **2001**, *1757*, 43–49. [CrossRef]
24. Blanck, G.; Cuisinier, O.; Masrouri, F. Soil treatment with organic non–traditional additives for the improvement of earthworks. *Acta Geotech.* **2014**, *9*, 1111–1122. [CrossRef]
25. Latifi, N.; Rashid, A.S.A.; Siddiqua, S.; Horpibulsuk, S. Micro-structural analysis of strength development in low- and high swelling clays stabilized with magnesium chloride solution—A green soil stabilizer. *Appl. Clay Sci.* **2015**, *118*, 195–206. [CrossRef]
26. Wang, F.; Xiang, W.; Corely, T.; Yeh, T.-C.J.; Yuan, Y. The influences of freeze-thaw cycles on the shear strength of expansives treated with ionic soil stabilizer. *Soil Mech. Found. Eng.* **2018**, *55*, 195–200. [CrossRef]

27. Yang, L.; Jiang, G.; Shi, Y.; Yang, X. Application of ionic liquid and polymeric ionic liquid as shale hydration inhibitors. *Energ. Fuels* **2017**, *31*, 4308–4317. [CrossRef]
28. Sameni, A.; Pourafshary, P.; Ghanbarzadeh, M.; Ayatollahi, S. Effect of nanoparticles on clay swelling and migration. *Egypt. J. Pet.* **2015**, *24*, 429–437. [CrossRef]
29. Luo, Z.; Wang, L.; Yu, P.; Chen, Z. Experimental study on the application of an ionic liquid as a shale inhibitor and inhibitive mechanism. *Appl. Clay Sci.* **2017**, *150*, 267–274. [CrossRef]
30. Mohammed, A.; Tariq, A.; Elhajj, H.; Walter, S.; Ibrahim, A. Unconventional Clay Control Alternative to Inorganic Compounds that Can Prevent Swelling and Reduce Friction in Underbalanced Drilling. In *SPE Middle East Oil and Gas Show and Conference*; Society of Petroleum Engineers: Dubai, UAE, 2019.
31. Chenevert, M.E. Shale alteration by water adsorption. *J. Pet. Tech.* **1970**, *22*, 1–141. [CrossRef]
32. Tandanand, S. Moisture adsorption rate and strength degradation of Illinois shales. In *The 26th US Symposium on Rock Mechanics (USRMS)*; American Rock Mechanics Association: Alexandria, VA, USA, 1985.
33. Gao, H.; Li, H.A. Pore structure characterization, permeability evaluation and enhanced gas recovery techniques of tight gas sandstones. *J. Nat. Gas Sci. Eng.* **2016**, *28*, 536–547. [CrossRef]
34. Sun, P.; Zhao, B.; Cao, H.; Wang, J.; Mo, D.; Zhang, S. Lab study on the effect of cation exchange capacity on slurry performance in slurry shields. *Adv. Civ. Eng.* **2018**. [CrossRef]
35. Sun, P.; Tian, M.; Cao, H.; Niu, L.; Zhang, S. Study on the mechanism of ENI action on preventing drilling fluid overflowing in HDD. *Tunn. Undergr. Sp. Tech.* **2018**, *77*, 94–102. [CrossRef]
36. Kawamoto, T.; Ichikawa, Y.; Kyoya, T. Deformation and fracturing behaviour of discontinuous rock mass and damage mechanics theory. *Int. J. Numer. Anal. Methods Géoméch.* **1988**, *12*, 1–30. [CrossRef]
37. Lu, Y.; Li, X.; Chan, A. Damage constitutive model of single flaw sandstone under freeze-thaw and load. *Cold Reg. Sci. Tech.* **2019**, *159*, 20–28. [CrossRef]
38. Kim, J.-S.; Lee, K.-S.; Cho, W.-J.; Choi, H.-J.; Cho, G.-C. A comparative evaluation of stress–strain and acoustic emission methods for quantitative damage assessments of brittle rock. *Rock Mech. Rock Eng.* **2015**, *48*, 495–508. [CrossRef]
39. Zhang, Z.; Zhang, R.; Xie, H.; Liu, J.; Were, P. Differences in the acoustic emission characteristics of rock salt compared with granite and marble during the damage evolution process. *Environ. Earth Sci.* **2015**, *73*, 6987–6999. [CrossRef]
40. Gu, X.B.; Wu, Q.H. The application of nonordinary, state-based peridynamic theory on the damage process of the rock-like materials. *Math. Probl. Eng.* **2016**, *2016*, 1–9. [CrossRef]
41. Jamshidi, A.; Zamanian, H.; Sahamieh, R.Z. The effect of density and porosity on the correlation between uniaxial compressive strength and P-wave velocity. *Rock Mech. Rock Eng.* **2018**, *51*, 1279–1286. [CrossRef]
42. Rajabzadeh, M.A.; Moosavinasab, Z.; Rakhshandehroo, G. Effects of rock classes and porosity on the relation between uniaxial compressive strength and some rock properties for carbonate rocks. *Rock Mech. Rock Eng.* **2012**, *45*, 113–122. [CrossRef]

© 2019 by the authors. Licensee MDPI, Basel, Switzerland. This article is an open access article distributed under the terms and conditions of the Creative Commons Attribution (CC BY) license (http://creativecommons.org/licenses/by/4.0/).

Article

Enhanced Tight Oil Recovery by Volume Fracturing in Chang 7 Reservoir: Experimental Study and Field Practice

Long Yu [1,2], Jinjie Wang [1,*], Chong Wang [3] and Daixin Chen [4]

1 Key Laboratory of Tectonics and Petroleum Resources (China University of Geosciences), Ministry of Education, Wuhan 430074, China; long.yu@ucalgary.ca
2 Department of Chemical Engineering, University of Calgary, Calgary, AB T2N1N4, Canada
3 Research Institute of Exploration & Development, Changqing Oilfield Company, PetroChina, Xi'an 710018, China; wch_cq@petrochina.com.cn
4 No. 12 Oil Production Plant, Changqing Oilfield Company, PetroChina, Xi'an 710018, China; chenzx_cq@petrochina.com.cn
* Correspondence: wangjinjie@cug.edu.cn

Received: 13 May 2019; Accepted: 19 June 2019; Published: 24 June 2019

Abstract: The Chang 7 reservoir in Changqing oilfield is rich in tight oil. However, due to the low formation permeability, it is very difficult to obtain economical oil production without stimulation treatments. Volume fracturing seems to be a more efficient tight oil recovery enhancement (EOR) method in Chang 7 pilot tests compared with conventional hydraulic fracturing. In this study, Chang 7 tight oil reservoir was first characterized by its geological property, hydrocarbon source rock distribution, and formation physiochemical property. Tight core flooding tests were then conducted to experimentally investigate the EOR ability of the volume fracturing technique. The field-scale practice was also demonstrated and analyzed. The results show that Chang 7 reservoir is favorable for the generation of a large amount of tight oil. Fractures created in tight cores can significantly improve the fluid flow conductivity and enhance the imbibition of displacing water, resulting in a greater tight oil recovery increment. Volume fracturing is an effective way to generate a larger number of fractures. Field application indicates that volume fracturing treatment can form a much greater reservoir stimulation volume. Daily oil production in the volume-fracturing-treated wells can be more than twice as high as that in the conventional-fracturing-treated wells.

Keywords: volume fracturing; Chang 7 reservoir; tight oil recovery; imbibition; oil production

1. Introduction

With the world's increasing need for energy and the gradual decline of conventional oil and gas production, attention has been turned toward unconventional reservoirs such as shale oil and shale gas [1]. Resources in tight reservoirs are substantial worldwide. Global gas reserves in shale reservoirs are estimated to be 716 trillion m^3 [2]. America has turned unconventional resource production into a commercial process. Tight sand production and shale gas account for about 30% of total gas production in America [3]. China has the largest natural gas resources in tight reservoirs or tight sand reservoirs [4,5]. However, although there is large resource potential in tight reservoirs, it is much more difficult to extract oil and gas from a tight formation than from conventional reservoirs, mainly due to the small pore size and the ultra-low permeability of the matrix [6,7].

Hydraulic fracturing combined with horizontal drilling has been widely used to enhance the production of oil and gas from tight sands or shale reservoirs [8,9]. Hydraulic fracturing improves hydrocarbon transport by increasing the formation's permeability and enhancing extraction. The main approach is to cut the low-permeability matrix into small pieces to increase the rock surfaces and connectivity. The fine induced fractures maximize the area of contact with the rocks, allowing hydrocarbon extraction from thousands of meters of rocks, rather than tens or hundreds of meters available with merely vertical wells [10,11]. Fracture networks connect each small piece of the matrix to the hydrocarbon flow path to the production well [12,13]. Many successful cases worldwide demonstrate that hydraulic fracturing enhances oil recovery in unconventional reservoirs [14,15]. Hydraulic fractures connecting with natural fracture systems ensure a more effective stimulation in fractured reservoirs, but the occurrence of arrest, diversion, and offset may inhibit the fracture growth and proppant placement. Blanton's [16] conducted laboratory hydraulic fracturing experiments using pre-fractured material to investigate the interactions between hydraulic fractures and natural fractures, aiming to provide information that helps to examine the conditions under which such inhabitations happen so that detrimental effects can be prevented. His results show that the morphology of hydraulic fracture is significantly affected by the pre-existing fractures, and hydraulic fractures are more likely to cross natural fractures only in high stresses and under high pre-existing fracture angles. Maity and Ciezobka [17] developed a novel image-processing workflow method to detect and analyze the proppant distribution in more than 200 through-fracture Permian Basin core samples to investigate proppant transport behavior in stimulated formation volumes during hydraulic fracturing treatment. The observations show that the presence of hydraulic fracture and of natural fracture and their interactions do not necessarily result in high proppant concentration in the cores, and stress contrast and lithology changes can govern the local in-depth proppant distribution. Their study provides a new method to systematically analyze core samples for the distributions of proppant and similar particles. Cipolla et al. [18] identified the benefits of tortuosity, multiple fractures, and the apparent activation of natural fractures and verified their potential using both fracture modeling and G-function analyses. Luo et al. [19] conducted a reservoir simulation using a geo-model to simulate the unique flow regime in multi-stage fractured horizontal wells compound formation linear flow. Results showed good agreement with field data of the Bakken formation.

In recent years, due to the rapid development of the tight oil/gas industry, the oil/gas recovery technique has made great progress. The appearance of volume fracturing has resulted in a great increase in tight oil and gas production [20]. The conventional hydraulic fracturing treatment aims to create a bi-wing open fracture in tight reservoir formations [21]. The fluid flowability is mainly affected by one fracture, but the fracture cannot improve the flowability of the total reservoir formation because, in the vertical direction of the fracture, the oil/gas still experiences a long-distance matrix flow in the low permeability porous media, as shown in Figure 1a. The single fracture limits the final oil/gas recovery efficiency, as the fluid flowability in the vertical direction is not increased [22,23]. To improve reservoir productivity, the fracture should be designed with a high fracture length and a high fracture conductivity [24]. The volume fracturing technique can generate single open fractures and can create and extend complex fracture networks [23,25]. In volume fracturing treatment, the reservoir formation was reconstructed and broken up through the fracturing method, forming several main fractures. By using the high flux, a large volume, and low viscosity fluids, together with the "staged and multi-cluster" perforation, secondary fractures can be generated in the main fractures. In the same way, second-order secondary fractures, and third-order secondary fractures, can also be created to form a complex fracture network in the formation [25,26], as shown in Figure 1b. The volume of the formation with fracture networks is called the stimulated reservoir volume (SRV) [27,28].

Fractures generated by volume fracturing are not only conventional open fractures, they also experience a shear/sliding dilation process [29], which leads to larger contact areas between the fracture faces and formation matrix. Due to the great number of created fractures, the complex fracture network and the large fracture-contact area, the oil/gas in the stimulated reservoir formations can flow along the shortest distance in any direction from the matrix to the fractures, significantly increasing the overall formation permeability [25,26,30]. Therefore, the reservoir formation can be effectively reconstructed, and reservoir productivity can be greatly improved.

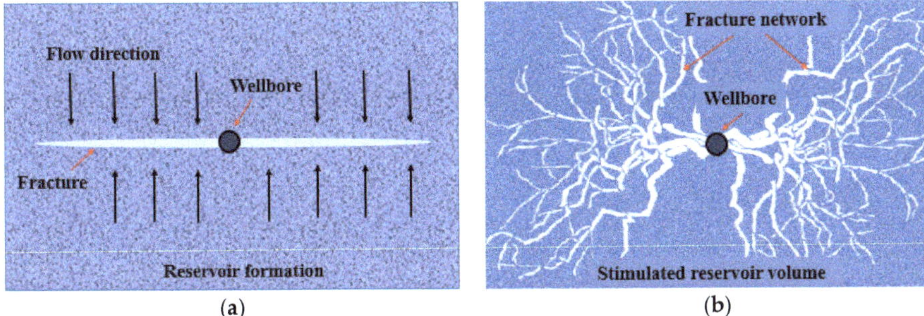

Figure 1. Schematic of conventional hydraulic fracture and fracture network by volume fracturing: (**a**) Conventional fracture, (**b**) fractures by volume fracturing treatment.

Both conventional fracturing and volume fracturing are sand fracturing techniques. According to some criteria, sand fracturing treatment with a fracturing fluid volume smaller than 1500 m^3 is called conventional fracturing, and that with a fracturing fluid volume larger than 1500 m^3 is called volume fracturing [31]. However, volume fracturing still has other characteristics in terms of operations. In volume fracturing, the fluid injection rate is relatively high, usually larger than 10 m^3/min, but it can be slightly smaller in some specific reservoirs. Volume fracturing is usually operated with a smaller proppant (70/100 mesh~40/70 mesh) and a lower sand ratio, with an average range of 3–5%. In some cases, the sand ratio can be around 10%, depending on the formation properties. Slickwater is mostly used as the fracturing fluid [32,33]. In some conventional fracturing theories, the injection of a significant volume of fracturing fluids with high pressures would cause a serious filter loss problem of the fluid in the matrix and forming shorter hydraulic fractures, resulting in formation damage and non-effective fracturing treatment. However, volume fracturing treatment always accompanies by injection of large volume fracturing fluids. First, this is because the segmented multi-cluster fracturing technique is usually employed in the volume fracturing treatment to form fracture networks [34–36]. In conventional staged fracturing treatment in horizontal wells, the single cluster perforation method is usually used, which can only support the low volume and low rate of fracturing fluid injection, not the large fluid injection required by the multi-cluster fracturing. Second, the volume fracturing is more suitable for and usually being applied in brittle formations, which are more easily to form a great number of complex fractures during the large volume injection fracturing treatments. Third, because of the low permeability of tight oil rocks, the filtration loss of fracturing fluids to the matrix is very small, most of the large volume fluids would flow into the numerous induced fractures or the natural microfractures, which could increase the stimulated reservoir volume. Study results from Li et al. [37] show that volume fracturing can significantly increase the production of tight oil reservoirs. A larger fractured volume and increased fracture networks could result in greater oil production. Chen et al. [31] compared the investment and benefit between conventional hydraulic fracturing and volume fracturing in the Southern Sichuan Basin and found that tested gas production yield from volume fracturing is about 2.12 times that of conventional fracturing technology. The wide application of volume fracturing in the field could reduce the circle returns and decrease total costs.

The Changqing oilfield is a regional oilfield affiliated with PetroChina Company Limited (PetroChina). In 2012, the oil-gas equivalent exceeded 45 million tons, making it the biggest field in China. The work area is located in the Ordos Basin, with an exploration area of 3.7 × 105 km^2. Chang 7 is a block rich in tight oil with estimated resources reaching up to two billion tons. Hydraulic fracturing has been tried for tight oil recovery in Chang 7. However, pilot tests and field data show that conventional hydraulic fracturing treatments cannot efficiently improve the oil production rate, whereas volume fracturing treatment in the Chang 7 tight oil formations can lead to a significant increase in the tight oil production and a great decrease in the water cut.

In this study, the geological and physical properties of Chang 7 tight oil reservoir were characterized first. The tight oil recovery ability of the volume fracturing treatment was investigated through core flow experiments and field scale practice. The results of this work may provide some instructions with the applicability of the volume fracturing technique in tight oil reservoirs for EOR.

2. Characterization of Chang 7 Tight Oil Reservoir

2.1. Geological Characteristics

Figure 2 shows the location and typical sedimentary facies of the Chang 7 reservoir, the Changqing oilfield. Chang 7 is the main tight oil reservoir in China and is located in the Ordos Basin. The Ordos Basin is a typical inland sedimentary basin with a multiphase craton in the southwest of North China. Delta facies and lacustrine facies were widely developed in the Ordos Basin. The delta front sand bodies, the gravity current settled sediments in semi-deep–deep lake faces, and the turbidity sand bodies are the most important reservoir sand bodies for Chang 7. The reservoir sand bodies are close to the hydrocarbon source rocks, possessing favorable conditions for the accumulation of tight oil.

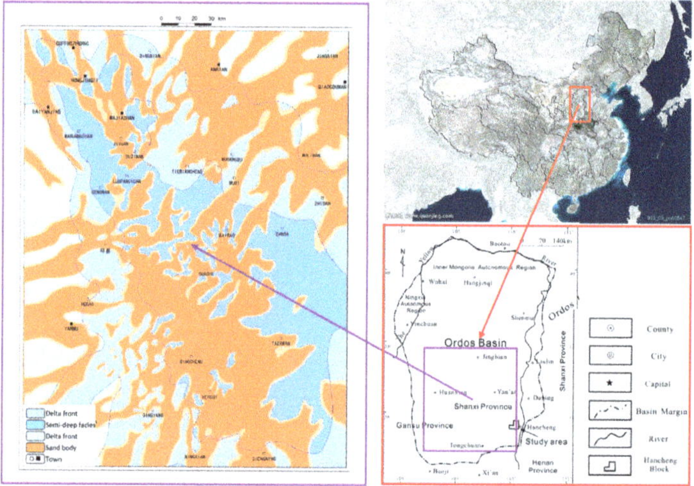

Figure 2. Location of the Ordos Basin together with typical sedimentary facies and sand body distributions of the Chang 7 tight oil reservoir, the Changqing oilfield.

2.2. Characteristics of Hydrocarbon Source Rocks

The hydrocarbon source rocks of Chang 7 are mainly composed of oil shale and dark mudstone. Figure 3 shows the isoline graphs of the thickness of the source rock distributions in Chang 7, the Ordos Basin. The oil shale, found largely at the bottom of the Chang 7 oil reservoir group, distributes in the NW–SE direction, reaching ZhiDan at the northeast, Majiatan at the northwest, Zhenyuan-Jingchuan at the southwest, and Yijun-Huangling at the southeast, and it is estimated that 37% of the total volume of source rocks in this area are oil shale, with an average thickness of 16 m and a total area of 3.25 × 104 km^2, as shown in Figure 3a. The hydrocarbon generation center for oil shale is at Jiyuan-Huachi-Zhengning. Compared with oil shale, the dark mudstone possesses a wider distribution. It reaches Ansai at the northeast, Majiatan at the northwest, Pingliang at the southwest, and Huanglong at the southeast, and the distribution of mudstone covers an area of 5.11 × 104 km^2 in the direction of NW–SE, with an average thickness of 17 m, as shown in Figure 3b. The mudstone occupies 63% of the total volume of source rocks.

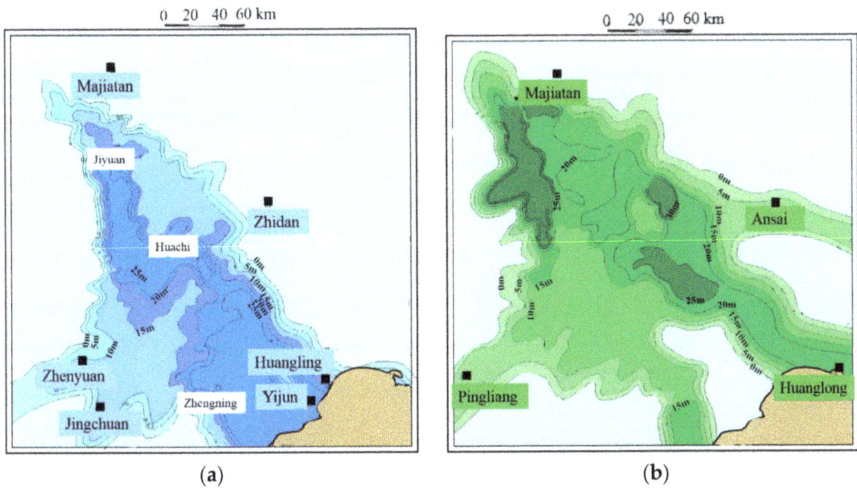

Figure 3. Isoline graph of the thickness of hydrocarbon source rock distributions in Chang 7, the Ordos Basin: (**a**) Oil shale, (**b**) dark mudstone.

The total organic carbon (TOC) is an important characterization parameter of the organic matter abundance in hydrocarbon source rocks. In this study, 123 core samples from Chang 7 were first collected, and a carbon and sulfur analyzer (CS analyzer, model: CS 230, LECO, St. Joseph, MO, USA) was used to measure the TOC in the hydrocarbon source rocks. The CS 230 analyzer can conveniently measure the TOC data with high accuracy and high stability, with a 1 ppm or 5% relative standard deviation (RSD) measurement accuracy. Hydrocarbon source rocks of Chang 7 are rich in organic matter. Organic geochemical data indicate that the residual organic carbon content in the hydrocarbon sources rocks mainly distributes in the range of 2–22%. In some source rocks, the residual organic carbon content can be as high as 30–40%. Figure 4 shows the distributions of the total organic carbon (TOC) for dark mudstone and oil shale in the Chang 7 reservoir. It can be seen that TOC for dark mudstone is smaller than 10, it mainly concentrates in the range of two to five. While for oil shale, the TOC has a wider range from 2 to 25. The average TOC for dark mudstone and oil shale is 5.8% and 13.8%, respectively.

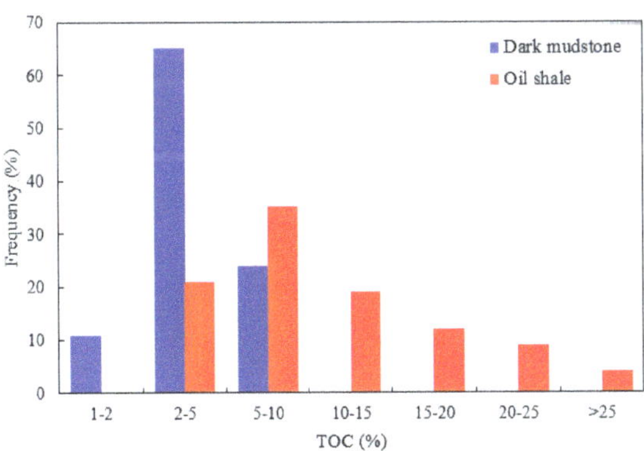

Figure 4. TOC distribution for source rocks in Chang 7 in the Ordos Basin.

The corresponding vitrinite reflectance (Ro) for the source rocks of the Chang 7 reservoir is between 0.85% and 1.15%, indicating a high intensity of hydrocarbon generation. It is estimated that the average intensity of hydrocarbon generation in Chang 7 is 495 × 10^4 km^2. The total amount of generated hydrocarbons in the source rocks can be as high as 2473.08 × 10^8 ton. The source rocks also experienced a strong hydrocarbon expulsion process, with a hydrocarbon expulsion intensity of 290 × 10^4 km^2, and a hydrocarbon expulsion amount of 1447.71 × 10^8 ton. It can be seen that the Chang 7 reservoir in the Changqing oilfield possesses a significant amount of favorable hydrocarbon source rocks, which ensures a high capacity for generating and producing oil.

2.3. Physical Properties of Reservoir Formations

Testing data collected from four pilot areas, Maling, Heshui, Jiyuan, and Ansai, in Chang 7 were used to characterize the physical properties of the tight oil reservoir formations. Figure 5 shows the location distribution of the four pilot areas. As can be seen, the pilot areas covered most of the typical formations in the Chang 7 tight oil reservoir, so the physical properties results measured in these areas can be used to characterize the overall physical properties for Chang 7.

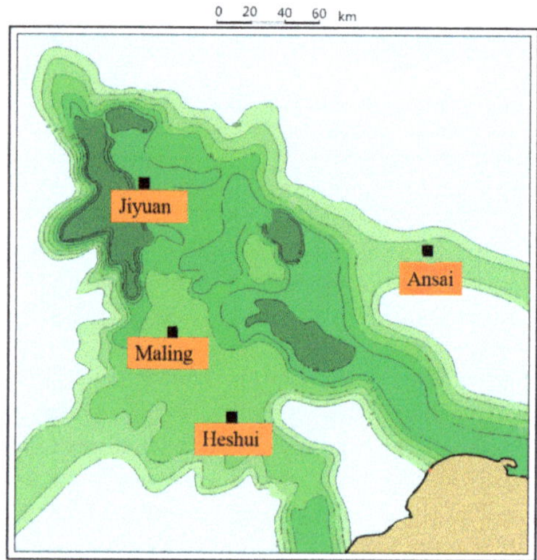

Figure 5. Locations of the four pilot areas in Chang 7 reservoir.

Table 1 gives the detrital components of the tight reservoir rocks of the four pilot areas in Chang 7. It can be seen that the tight sandstones are mainly composed of quartz, feldspar, and lithic fragments, which occupy 77.3% of the total composition analyzed through a number of rock samples. Therefore, the formation rocks possess a high brittleness and tend to form complex fractures during fracturing treatments.

Table 1. Detrital components for the four pilot areas in Chang 7.

Pilot Area	Number of Sample	Detrital Composition [%]			
		Quartz	Feldspar	Lithic Fragment	Others
Maling	68	40	18.4	18.4	23.2
Heshui	66	41.2	20.8	16.9	21.1
Jiyuanan	123	25	38.9	12.3	23.8
Ansai	31	26.8	40.6	11.4	21.2
Sum	288	32.5	30.1	14.7	22.7

To obtain the porosity and permeability distributions in Chang 7 reservoir, a number of core samples from the formation layers of four blocks of Maling, Heshui, Jiyuan, Ansai were collected and the porosity and permeability of the cores were measured. The core data, together with the reservoir well logging results, were used to draw the porosity and permeability maps. Table 2 shows the coring details, Figure 6 shows the porosity distributions for the blocks of Maling, Heshui, Jiyuan and Ansai, Figure 7 shows the permeability results. The dots and lines in the figures are the existing wells.

Table 2. The coring details for the analysis of porosity and permeability in Chang 7 reservoir.

Blocks	Formation Layer	Number of Core Samples	Average Porosity [%]	Average Permeability [mD]
Maling	Chang 7_2	2234	9.1	0.13
Heshui	Chang 7_1	2575	8.9	0.14
Jiyuan	Chang 7_2	3455	8.9	0.12
Ansai	Chang 7_3	4067	8.1	0.11

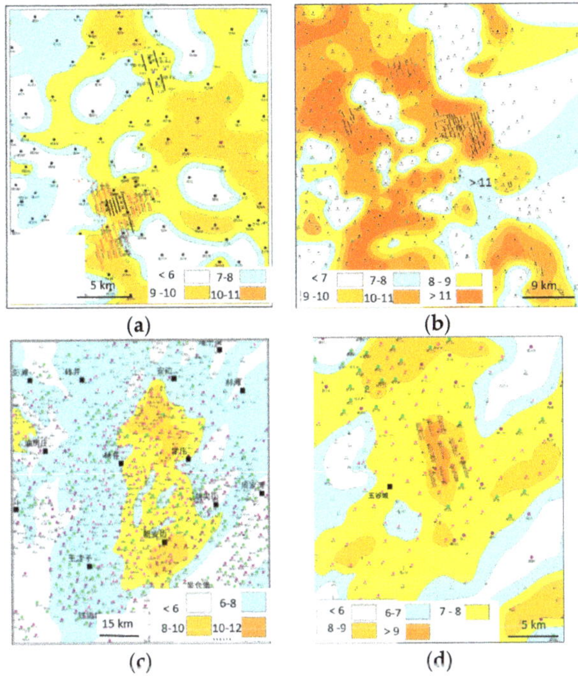

Figure 6. Porosity (%) distributions of the four pilot areas in Chang 7: (**a**) Maling, (**b**) Heshui, (**c**) Jiyuan, (**d**) Ansai.

Porosities of the reservoir formations in Maling, Heshui, Jiyuan, and Ansai mainly distributes in the range of 6–11%, 7–11%, 6–10%, and 6–9%, respectively. Most of the reservoir formation permeabilities are smaller than 0.2 mD (Figure 7a–c). In Ansai, the formation permeability can be much smaller, as shown in Figure 7d. The average porosity of Chang 7, as calculated based on the overall core analysis, is around 8–9%, and the average permeability is about 0.11–0.14 mD. As can be seen, the Chang 7 reservoir possesses very low porosity and permeability, which are unfavorable for oil production.

The micropore structure of the tight sandstones in the Chang 7 tight oil reservoir was measured using X-ray computed tomography (X-CT). X-CT is a widely used and reliable technology that can rapidly measure the pore structures of tight cores with high accuracy, and the measurement scale of CT ranges from nm to μm level. Figure 8 shows the variation in pore volume at different pore radii. The results were obtained based on an X-CT analysis of 52 cores from the Chang 7 reservoir. The pore radius of the Chang 7 tight oil reservoir mainly distributes in the range of 2–10 μm. Pores with a radius of 2–10 μm occupy 76.8% of the total pore volume in the reservoir and thus are the main storage space for the tight oil.

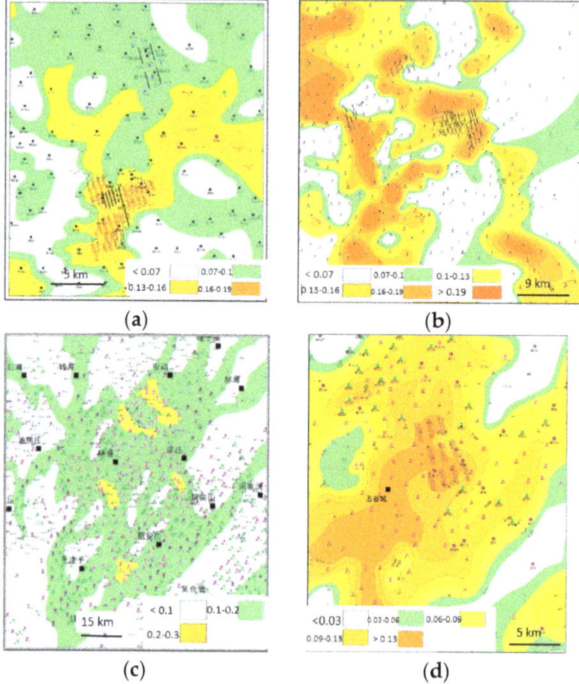

Figure 7. Permeability (mD) distributions of the four pilot areas in Chang 7: (**a**) Maling, (**b**) Heshui, (**c**) Jiyuan, (**d**) Ansai.

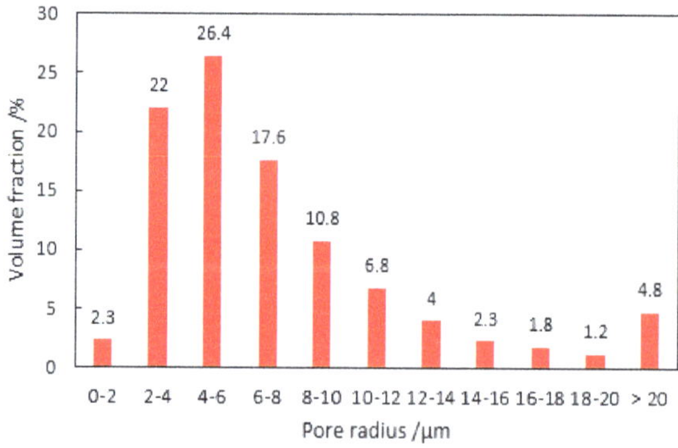

Figure 8. Variation of pore volumes at different pore radius for the tight rocks of Chang 7.

Figure 9 shows the coordination numbers of the pores for the Chang 7 tight oil reservoir. It can be seen that the overall coordination number for the pores is small, mainly concentrating in 1–3. Coordination number characterizes the number of pore throats that connect to one pore body. The smaller coordination number means poorer pore connectivity because, in such case, the pores were connected with fewer throats, which would limit the fluid flowability in the tight porous media. Although most of the pore body radius of Chang 7 reservoir are between 2 μm and 10 μm, and the pore throat radius could be much smaller. Results in Figure 8 indicates that it is difficult for the tight oils flowing in the porous media, and thus uneconomic to recover them from the reservoir without further stimulation treatment. To acquire economically feasible tight oils, hydraulic fracturing techniques should be utilized.

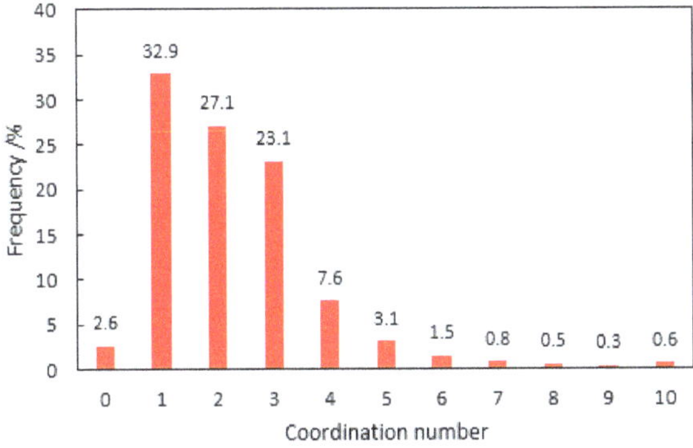

Figure 9. Coordination number for the pores of Chang 7 tight oil reservoir.

3. Experimental Section for Tight Oil Recovery in Pore-Fracture Dual Media

3.1. Flow Conductivity of Pore-Fracture Media and Its Effects on Tight Oil Recovery

The flow conductivity in a pore-fracture dual medium can be much higher than that in single pore media due to high permeability fractures. In order to investigate the flow behaviors of oil and water in the dual medium, relative permeability measurements were conducted through core flow tests. Because cores with fractures are easily broken, it is difficult to obtain them from the reservoir formations directly. In this study, the unbroken cores from Chang 7 formations, 2.5 cm in diameter and 6 cm in length, were compressed using the tri-axial compression method to create different microfractures. During the fracture creating process, we used the trial and error method to make fractured cores which possess appropriate initial permeabilities until the required cores were successfully achieved. The successfully made cores would be then used for the relative permeability tests. The relative permeability of oil and water were measured using the un-steady state method. The change of cumulative production of oil and water with time was recorded carefully to accurately calculate the relative permeability data. Figure 10 shows four typical results for the relative permeability variations of oil and water in the cores with different fractures. Before fracturing, the cores possess similar initial permeability (k_i). The core permeabilities after fracturing (k_f) are mainly determined by the developed fractures. The after-fracturing permeabilities of the core sample (Figure 10a,b) do not increase very much compared to the initial permeabilities. For core sample (Figure 10a), the permeability increases from 0.08 mD to 0.19 mD, and for core sample (Figure 10b), it increases from 0.14 mD to 0.36 mD. In such cases, the fractures created by the tri-axial compression method in the cores are more likely the pre-existing microfractures in formations. The fractures possess a small size, small number and have limited effects on the fluid flow conductivity increase in the cores. The permeabilities of the core sample (Figure 10c,d) have a significant increase after fracturing, the measured k_f can be as high as 5.48 mD and 12.14 mD, respectively. In such cases, the created fractures in the cores are more likely the produced fractures by fracturing treatment in formations. Such fractures were more complex and with a larger number and larger size than the fractures in core sample (Figure 10a) and core sample (Figure 10b).

As can be seen in Figure 10, before fracturing, the flow of oil and water occurs in the matrix of the cores. The relative permeability of oil first decreases rapidly with water saturation and then declines slowly. The relative permeability of water increases gradually. The cores possess a wide two-phase flow region. After fracturing, due to the developed microfractures, the flow conductivity in all of the cores has a great increment for both oil and water. For cores with slightly increased permeability, sometimes two to three times higher than that before fracturing (Figure 10a,b), the variations of relatively permeabilities for oil and water are similar to that in the un-fractured cores. The relative permeability of oil declines, and the relative permeability of water inclines gradually. The intersection of the two-phase relative permeability curves is higher in the fractured cores, indicating a greater increase in fluid flowability after fracturing. In addition, the fractured cores still have a wide two-phase flow region. If the permeabilities of the cores after fracturing are much higher than that before fracturing, as shown in Figure 10c,d, there is no doubt that flow conductivity in the cores can be improved greatly. However, the relative permeabilities for the oil phase and the water phase change drastically with water saturation in these fractured cores. The two-phase flow region becomes much smaller than the un-fractured cores and the slightly fractured cores. In such cases, a more rapid water breakthrough will occur through these fractures in the cores when flooded with water. Therefore, in order to efficiently increase oil recovery in tight oil formations through fracturing treatments, such fractures with an appropriate flow conductivity should be created. The developed fractures in formations could enhance the fluid flowability.

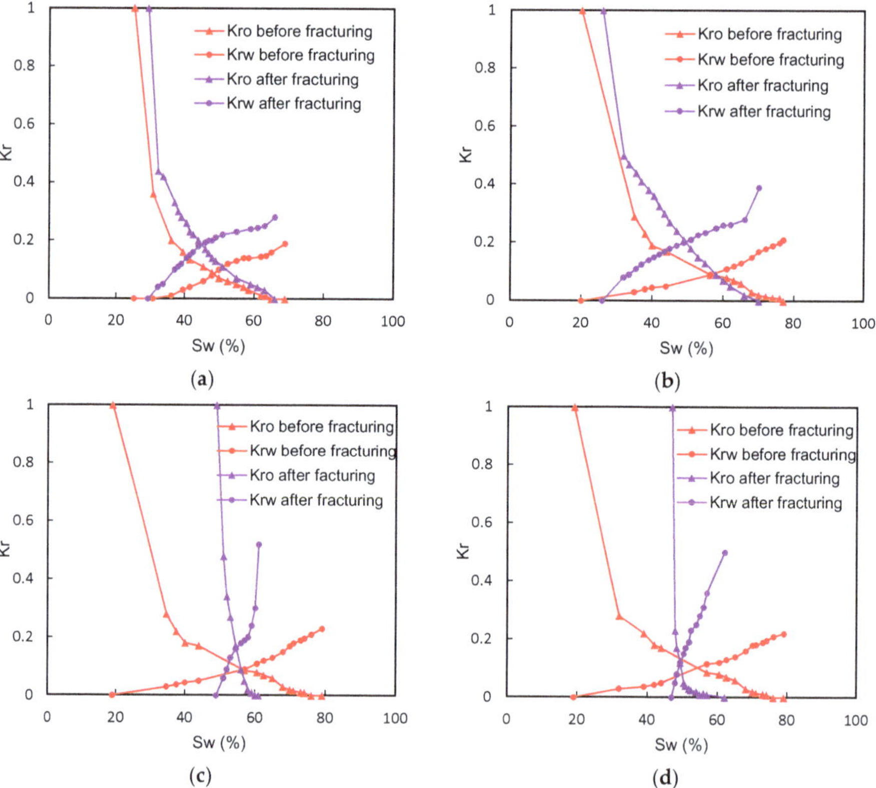

Figure 10. Relative permeabilities of oil and water in the cores before and after fracturing: (**a**) $k_i = 0.08$ mD, $k_f = 0.19$ mD, (**b**) $k_i = 0.14$ mD, $k_f = 0.36$ mD, (**c**) $k_i = 0.07$ mD, $k_f = 5.48$ mD, (**d**) $k_i = 0.07$ mD, $k_f = 12.14$ mD.

Figure 11 shows the variations of tight oil recoveries in the fractured cores at different permeability ratios of the after-fracturing core permeability to the before-fracturing core permeability (k_f/k_i).

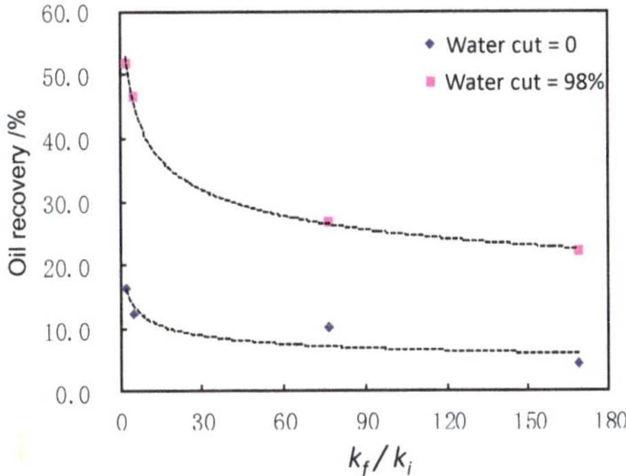

Figure 11. Oil recovery variations with different permeability ratios at different water cuts.

It can be seen that both of the tight oil recoveries of the cores decrease with an increase of the permeability ratio k_f/k_i. Before water breakthrough (water cut = 0), the decrease of oil recovery with permeability ratio is small. However, when water cut = 98%, the tight oil recovery decreases significantly as the permeability ratio increases. This is because the created fractures in the cores possess a higher permeability than the matrix, and due to the small size of the cores, some of the fractures may form through the whole cores from the inlet to the outlet. These fractures could help the oil in the matrix to flow out more easily and thus can improve the oil production rate. However, due to the high conductivity, once the water breaks through from these very high permeability fractures, especially from those that connect the core inlet to the core outlet, the following displacing fluid would be more likely to flow through these highly permeable water channels, rather than flow into the matrix to displace oil. This water channeling problem could be more severe as the permeability of the fractures becomes greater, leaving more residual oil in the tight cores [38]. In the field scale, because of the long distance between the injection well and the production well and the appropriate operations, most of the fractures are unlikely to break through the wells to lead to the water channeling problem. Therefore, fracturing treatment in field can usually result in high oil production rate like the core flooding but may rarely cause the water channeling problem as in the core experiments.

3.2. Effect of Imbibition on Oil Recovery in Fractured Tight Oil Reservoir

In conventional oil reservoirs, the oil is flooded out from the pore spaces by the displacing fluids. The oil recovery is improved mainly by the displacement effect. However, in tight oil reservoirs, because of the ultra-low permeability, it is much more difficult for the tight oil to be displaced out by the injected water than the conventional oil. In such reservoirs, apart from the displacement effect, the imbibition of water in the pore spaces plays an important role in tight oil recovery improvement. Fracturing treatment can improve the oil production rate in tight reservoirs. The matrix of reservoir formations provides spaces for tight oil storage, and the created fractures mainly provide highly conductive flow channels for the oil. In the early recovery stage, most of the produced oil comes from the fractures. As the oil in the fractures flows out, the following oil would be mainly from the matrix. In this stage, water in the fractures imbibes into the low permeability matrix to displace tight oil out from the pore spaces. Such an imbibition effect is important for the oil recovery in the fracturing-treatment tight reservoirs, especially when the reservoir is in the mid or late development stages.

Figure 12 shows a detailed schematic of the imbibition effect on oil recovery in a tight porous medium with fractures. Imbibition in pore-fracture dual media can be described as the wetting phase (water) in fractures spontaneously imbibed into the matrix of the tight porous media under the capillary force, replacing the non-wetting phase (oil) in the tight pore spaces. The replaced oil will then enter the fracture systems and finally flow into the wellbore. The complex fractures in tight porous media divide the total matrix into a number of smaller pieces, which provides more interaction areas for the water and the oil. Therefore, more displacing water can invade the matrix to replace the oil inside. The imbibition effect would thus be enhanced. As the oil flows out from the matrix, the dense fractures can also provide more flow channels for the replaced oil, which increases the possibility of further movement of the oil.

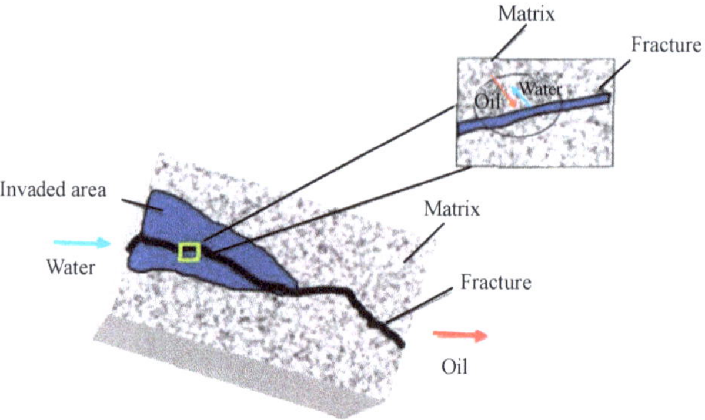

Figure 12. Schematic of the imbibition effect on oil recovery in tight porous media with fractures.

Figure 13 shows a comparison test of the water imbibition rate in tight cores with and without fractures. The permeability of the matrix of the two tight cores is 0.08 mD. Fractures in the cores were created using a tri-axial compression apparatus.

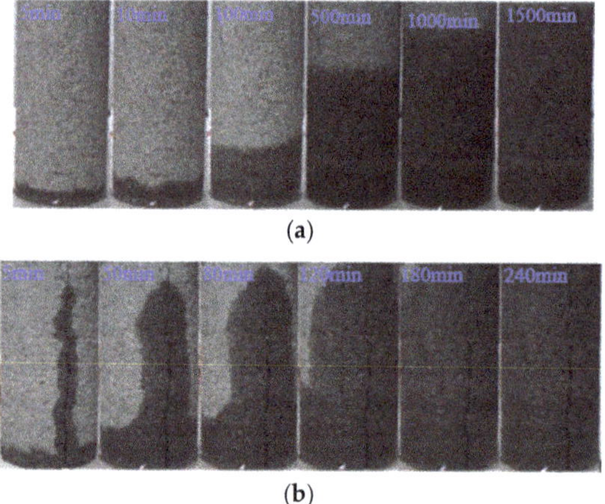

Figure 13. Images of imbibition rate of water in tight cores: (**a**) Core without fractures, (**b**) core with fractures.

As can be seen in Figure 13, the imbibition process in the tight core with fractures is much faster than that in the core without fractures. In the core without fractures, the water was imbibed into the matrix slowly from the bottom. In the core with fractures, the water tended to first enter into the fractures rapidly, and the water then begins to be imbibed into the nearby matrix along the fractures. After about 240 min, the core with fractures can be completely occupied by the imbibed water, more than five times faster than the core without fractures. The fractures in the tight cores can significantly increase the contact areas for the water and the core matrix and thus can improve the imbibition rate in the cores remarkably. Provided that the cores were saturated with oil in advance, a higher oil producing rate enhanced by the imbibition effect can be expected in the fractured cores.

Capillary pressure is the main driving force of the imbibition effect in tight porous media [39,40]. As the capillary pressure is inversely proportional to the capillary radius, the smaller capillary possesses greater capillary pressure in it and thus has a stronger imbibition effect. The average pore diameter for tight porous media can be several to tens of nanometers, which ensures a high capillary force in those pores. To have an effective oil recovery in the fracturing treatment of tight oil reservoirs, the capillary pressure or the imbibition should be high enough to overcome the end effect between the fractures and the matrix. Another thing one should be concerned with is that a smaller pore radius is not always helpful for the imbibition-enhanced oil recovery. In order to effectively replace oil with water in tight porous media, the pore radius should be larger than the thickness of the water adsorption layer on the pore walls.

During the water flooding process in tight oil reservoirs, apart from the displacement effect, oil recovery can also be enhanced by the imbibition effect, as discussed above. To determine the oil recovery factors of the displacement effect and the imbibition effect, flooding tests for Chang 7 tight cores with different permeabilities were performed, and nuclear magnetic resonance (NMR) technology was used to analyze the different oil recoveries.

Figure 14 shows a typical frequency spectrum of the transversal relation time (T_2) measured by NMR for a tight oil core at different oil saturations.

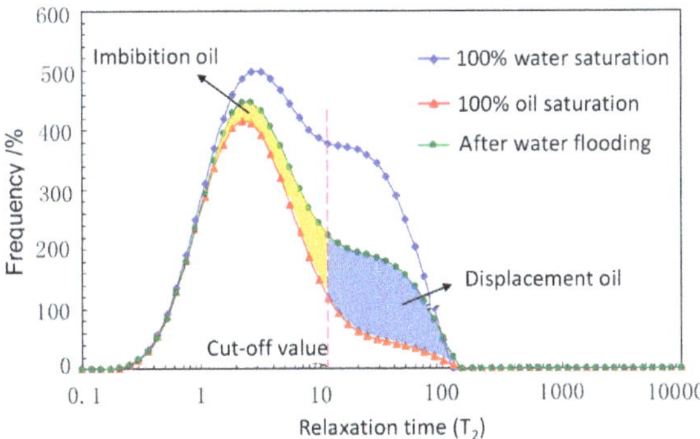

Figure 14. Frequency spectrum of relation time for tight oil core at different oil saturations.

In Figure 14, the area between the T_2 frequency at 100% water saturation (blue curve) and the T_2 frequency at 100% oil saturation (red curve) characterizes the total oil saturation in the tight core. The area between the T_2 frequency after water flooding (green curve) and the T_2 frequency at 100% oil saturation characterizes the total movable oil in the tight core. As can be seen, in tight oil reservoirs, due to the low permeability and poor pore connectivity, a large amount of oil cannot be moved and produced and will remain in the reservoir as residual oil. The water flooding process in tight oil reservoirs is designed to recover the movable oil as much as possible. During the water flooding process, the movable oil is produced both by the water imbibition effect and the water displacement effect. The relaxation time cut-off value was used to characterize the limitation of movable oil recovered by the two different effects [41–43]. As shown in Figure 14, for oil recovered by the imbibition effect (yellow area), the corresponding T_2 is smaller than the cut-off value, and the corresponding T_2 is larger than cut-off value.

Based on the NMR analysis of the water flooding tests in the tight cores, oil recoveries improved by the imbibition effect and the displacement effect were obtained. Table 3 shows the detailed movable oil recovery results for the imbibition effect and the displacement effect in 17 tight oil cores from the Chang 7 reservoir.

Table 3. Movable oil recoveries affected by imbibition and displacement in Chang 7 tight cores.

Core Number	Permeability [mD]	Porosity [%]	Total Movable Oil Saturation [%]	Recovered Movable Oil Saturation [%]		Recovery of Movable Oil [%]	
				Imbibition Effect	Displacement Effect	Imbibition Effect	Displacement Effect
1	0.014	7.5	24.42	9.81	14.61	40.17	59.83
2	0.014	6.7	25.45	13.06	12.39	51.32	48.68
3	0.015	7.1	28.42	12.77	15.65	44.93	55.07
4	0.034	10.1	34.21	13.21	21	38.61	61.39
5	0.038	10.7	34.16	10.77	23.39	31.53	68.47
6	0.044	9.9	35.31	13.23	22.08	37.47	62.53
7	0.045	10.2	37.06	13.48	23.58	36.37	63.63
8	0.058	10.5	35.31	11.01	24.3	31.18	68.82
9	0.065	10.3	32.7	10.12	22.58	26.00	58.02
10	0.066	10	38.92	10.8	28.12	28.18	73.38
11	0.067	11.9	38.32	9.46	28.86	24.69	75.31
12	0.096	11.3	36.87	6.67	30.2	18.09	81.91
13	0.102	11.1	37.72	5.29	32.43	14.02	85.98
14	0.104	10.7	34.81	3.23	31.58	9.28	90.72
15	0.108	11.7	39.04	5.22	33.82	13.37	86.63
16	0.135	10.9	39.79	2.12	37.67	5.33	94.67
17	0.143	8.6	36.01	4.59	31.42	12.75	87.25

As we can see in Table 3, because of the low permeability, the tight cores obtained from the Chang 7 reservoir possess a low movable oil saturation (<40%). During the water flooding process, the imbibition effect could have a significant contribution to the movable oil recovery in the tight cores, as about 5.33–51.32% of the movable oil recovery was improved by the water imbibition. It can also be seen in Table 3 that the imbibition effect in tight cores decreases with the increase in core permeability. In the cores with lower permeabilities, e.g., where $k < 0.06$ mD, a large portion of movable oil (around 30–50%) could be recovered by the imbibition effect. In cores with higher permeabilities, the water displacement takes a much more predominant effect.

The measurements and experimental tests in this study can provide instructive data to further reservoir simulation and optimization operations. The permeability, porosity and pore structure distribution data obtained from thousands of core sample analysis could help to enhance the reliability of the accurate description for the physical properties of reservoir formations. The relative permeability tests provided useful data for the analysis of flow behavior of tight oil and water in pore-fracture dual media. The EOR results from core flooding experiments could give instructions to the pilot operations during the fracturing treatments, i.e., when enhancing oil recovery in tight oil reservoirs using fracturing treatment, on one hand, the developed high permeability fractures could increase the oil production rate, on the other hand, such existing high permeability channels could lead to a higher water cut and a smaller final oil recovery factor. Thus, creating proper fractures considering both the production rate and final oil recovery is of great significance to the application of fracturing in tight oil reservoirs.

4. Field Practice of Volume Fracturing Treatment

4.1. Field Development Stages in the Chang 7 Reservoir

The field development in the Chang 7 tight oil reservoir has experienced four stages. The first is the directional well development stage. Directional well tests were conducted from 2010 to 2011. During this stage, water injection accompanied by multistage sand fracturing in directional wells was performed. The average amount of ceramic proppant injected into each well was 32.1 m^3. The sand ratio in the fracturing fluid was 29.8%, and the fluid injection rate was 2.0 m^3/min. The production rate in the treated directional wells was low, about 1.5 ton/day in every well, indicating little improvement to the tight oil recovery.

To solve the problem of the low production rate in the directional wells, in 2011, conventional fracturing treatment in horizontal wells was tried in the Chang 7 reservoir. The fluid injection rate of the conventional fracturing was smaller than 6 m^3/min. The production results in six treated horizontal wells showed that a significant oil production rate improvement can be obtained in the early well production stage, reaching about 5.3 ton/day in every well. However, the wells were soon watered out and oil production rate dropped rapidly.

In 2012, the volume fracturing pilot test in the Chang 7 reservoir was carried out. Twenty-four horizontal wells were treated by volume fracturing during this stage. Compared with conventional fracturing, the fluid injection rate of volume fracturing was higher than 6 m^3/min, with a larger injected fracturing fluid volume. After volume fracturing treatment, the oil production rate in horizontal wells can be markedly increased to as high as 10 ton/day, almost twice as high compared with the conventional fracturing treatment. In Chang 7 reservoir, a segmented multi-cluster fracturing with fast-drilling bridge plugs technique was applied during the volume fracturing treatment [38], as shown in Figure 15. The long horizontal well was divided into several intervals by bridge plugs, and three to five clusters perforations were performed in every interval. The multi-cluster perforations can support the large volume fluid injection which is required to form a great number of complex fractures. In addition, due to the shorter inter-fracture distance in the multi-cluster fracturing, the interference between them can generate more complex fracture networks [25]. In Chang 7 reservoir, the formation rocks possess a high brittleness as the quartz, feldspar and lithic fragment composed over 70% of the rocks, which are more easily to form a great number of complex fractures during the large volume injection fracturing treatments.

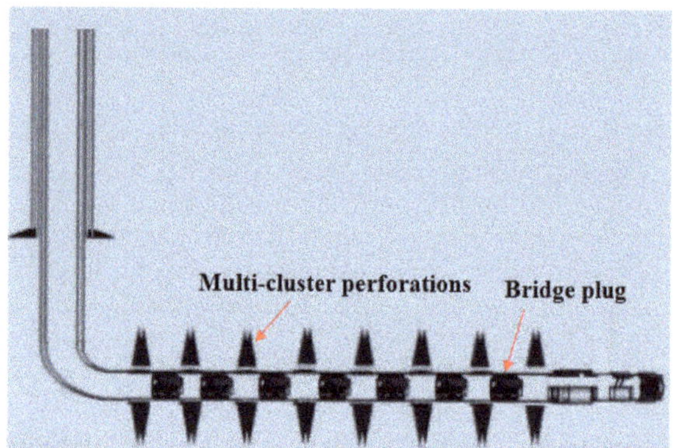

Figure 15. Schematic of the segmented multi-cluster fracturing technique with bridge plugs.

Considering the significant promotion of oil production with the pilot volume fracturing treatment, a large-scale field application of hydraulic volume fracturing in four blocks in the Chang 7 reservoir was launched from 2013 to 2016. By the end of 2016, 414 wells in Chang 7 had been treated with volume fracturing. The average oil production rate of the treated wells was 9.6 ton/day per well for the first three months. So far, the total tight oil production in these wells has been 3.203×10^6 ton.

4.2. Tight Oil Production in Volume-Fracturing-Treated Wells

Table 4 shows a set of well stimulation results by the volume fracturing treatments in the Chang 7 tight oil reservoir. Fractures parameters were obtained through the application of microseismic monitoring technology. SRV_f represents stimulated reservoir volume by conventional fracturing treatment, and $SRVv_f$ represents stimulated reservoir volume by volume fracturing treatment. It can be seen that the volume fracturing treatments in the wells possessed a higher fluid injection rate, a lower sand ratio, and a larger reservoir stimulation volume than the conventional fracturing. The average injection rate of volume fracturing in 12 treated wells was about 8.8 m^3/min, almost four times that of the conventional fracturing. The sand ratios, which ranged from 8.5 to 14.3%, were less than half those in conventional fracturing. The volume fracturing treatment in wells formed favorable fractures in the formation. The average half-length, width, and height of the fractures created by volume fracturing treatments were all larger than those in conventional fracturing treatments, indicating an improvement in the fluid flow in the fractures. Due to the proper parameter design, such as large sand and fluid injection volume, low sand ratio, high injection rate, and so on, and the appropriate well operations, volume fracturing could lead to a larger fracture distribution around the treated wells. The volume fracturing treatment design overall resulted in the greater SRV in the reservoir, i.e., the largest SRV_{vf} in the treated wells was 4.9 times greater than the SRV_f, which also benefits the fluid flow and the final oil production.

Table 4. Reservoir stimulation results of volume fracturing treatments in Chang 7.

Fracturing Treatment	Well No.	Field Operation Parameters					Fracture Parameters			SRV [10^4 m^3]	SRV_{vf}/SRV_f
		Number of Well Intervals/Number of Perforation Clusters	Sand Volume [m^3]	Sand Ratio [%]	Injection Rate [m^3/min]	Total Injected Volume [m^3]	Fracture Half-Length [m]	Fracture Width [m]	Fracture Height [m]		
Volume fracturing	1	13/26	514.8	9.5	6.2	7475	257	110	69	387.1	2.7
	2	13/26	520	9.3	6.3	7800	272	96	78	402.8	2.8
	3	11/22	496.1	10.6	6.0	6556	401	38	58	176.8	1.2
	4	11/22	438.9	8.5	6.0	7656	326	69	51	229.4	1.6
	5	11/22	482.9	10.7	6.0	5896	371	62	52	240.4	1.7
	6	9/44	1146.6	13.8	15.0	9135	386	162	41	515.3	3.6
	7	8/39	1016	14.3	15.0	10248	336	128	37	314.0	2.2
	8	4/11	259.2	11.0	10.0	2148	302	71	68	289.1	2.0
	9	4/12	312.4	11.0	9.4	3352	291	68	69	270.3	1.9
	10	8/32	475.2	12.9	7.7	4152	337	110	96	711.7	4.9
	11	6/24	520.8	12.4	8.0	4680	244	136	38	252	1.8
	12	5/21	377.5	13.0	8.0	3910	160	120	48	183.4	1.3
Conventional fracturing	13	8/16	184	29.0	2.2	1223	160	90	50	144	1.0

SRV—stimulated reservoir volume, SRV_{vf}—stimulated reservoir volume by volume fracturing treatment, SRV_f—stimulated reservoir volume by conventional fracturing treatment.

Volume fracturing is always accompanied by large fluid injection volume, it also has a characteristic of low backflow rate. During the fracturing treatment, most of the fracturing fluid was retained in the formations, which could increase the formation pressures. In Chang 7 reservoir, the average formation pressure, P_i, is 15.8 MPa. Bai et al. [35] investigated the variation of fracture net pressure with the change of injection rate in formation layers with different thickness during hydraulic fracturing treatment in Chang 7 reservoir. Figure 16 summarized characteristics of the net pressure variation in formation fractures at different injection rates. As we can see, increasing the injection rate can improve the net pressure in the formed fractures.

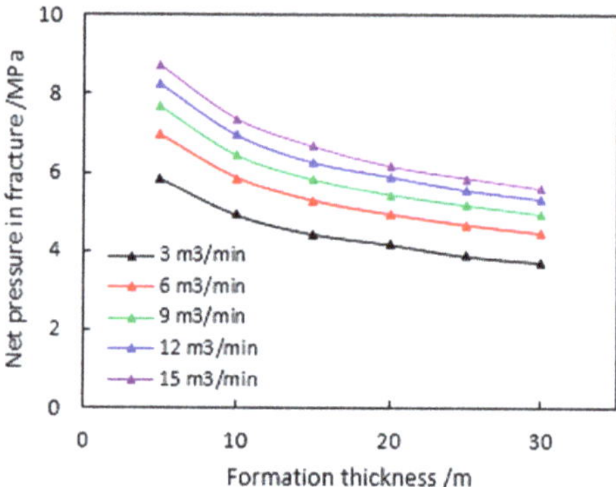

Figure 16. Variation of net pressures in formed fractures with the change of injection rate in formation layers with different thickness during fracturing treatments.

After injecting large volume fracturing fluid, the average formation pressures, P, in different blocks increased. Table 5 gives the formation pressure data during the volume fracturing treatments in some of the pilot areas. The pressures were obtained based on both the direct pressure measurements and simulation results.

Table 5. Formation pressures during volume fracturing treatments in Chang 7.

Pilot Areas	Number of Wells	Average Length of Horizontal Well [m]	Fluid Injection Volume per Well [m^3]	Fluid Backflow Volume per Well [m^3]	Backflow Rate [%]	Fluid Volume Retained in Formation [m^3]	P/Pi [%]
Jiyuan	173	789	5463	1672	30.6	3791	119
Heshui	56	976	5805	1623	28.0	4182	119
Maling	39	1264	9294	3263	35.1	6031	127

In 2011, six horizontal wells in Jiyuan pilot areas were treated by conventional fracturing with low fluid injection volume and low injection rate. The average oil production rate in the wells was 5.3 ton/day, and the production wells were quickly flooded after a few months. In 2012, 24 horizontal wells in Jiyuan pilot areas were treated by volume fracturing. After the treatment, the average oil production rate in the wells reached to 12 ton/day, twice larger than that of the conventional fracturing treatment. After 2012, a number of horizontal wells in Jiyuan, Heshui, and Maling were treated with volume fracturing. The average daily oil rate after one-year production in the treated wells were still higher than eight tons, which indicating a slowly decrease in oil productivity. Table 6 shows a summary of the decline rate of oil production in conventional fracturing and volume fracturing horizontal wells. As we can see, after one-year production, the average oil decline rate in volume fracturing wells is lower than that in conventional fracturing wells. Figure 17 shows a typical comparison of the one-year tight oil production results in two horizontal wells treated by volume fracturing and conventional fracturing in the Chang 7 reservoir. As can be seen, the volume fracturing treatment in horizontal wells can significantly increase tight oil production. In the first month, the average daily oil production of the volume-fracturing-treated wells was more than twice that of the conventional-fracturing-treated wells. The daily oil production decreased with the production time in both types of treated wells. Compared with conventional fracturing, the tight oil production in the wells treated by volume fracturing had a slower decreasing rate. By the end of one year, the daily oil production of the volume-fracturing-treated wells had declined from 12.3 to 7.6 ton/day, about a 38% decrease. The daily oil production of the conventional-fracturing-treated wells had a more remarkable decrease of 55%, from 5.15 to 2.3 ton/day. The one-year-accumulated tight oil production of the volume-fracturing-treated wells was as high as 3500 t, about 2.3 times larger than that of the conventional-fracturing-treated wells, as shown in Figure 17b.

Table 6. Decline of oil production in conventional fracturing and volume fracturing horizontal wells.

Fracturing Treatment	Number of Wells	Pilot Areas	Well Length [m]	Sand Volume per Well [m^3]	Injection Rate [m^3/min]	Fluid Injection Volume per Well [m^3]	One-year oil Production Decline Rate [%]
Conventional fracturing	5	Jiyuan	<1000	174.2	2.1	974.2	57.3
Volume fracturing	37	Maling, Heshui, Jiyuan	1500	763	7.1	12511	34.1
	13	Heshui, Jiyuan	1000–1400	715	6.8	8791	40.5
	2	Jiyuan	<1000	494.8	6.0	5241	37.8

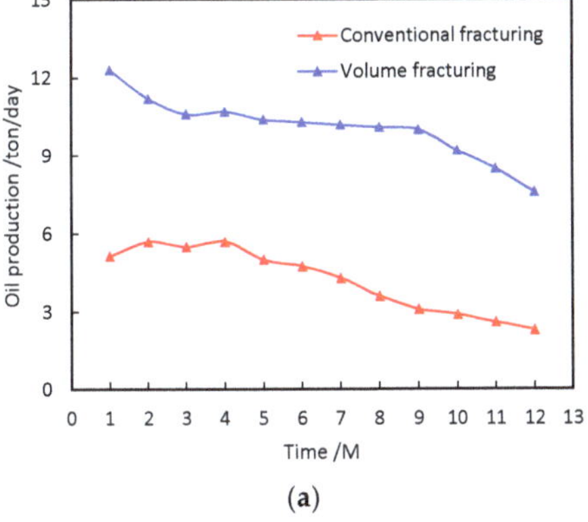

(a)

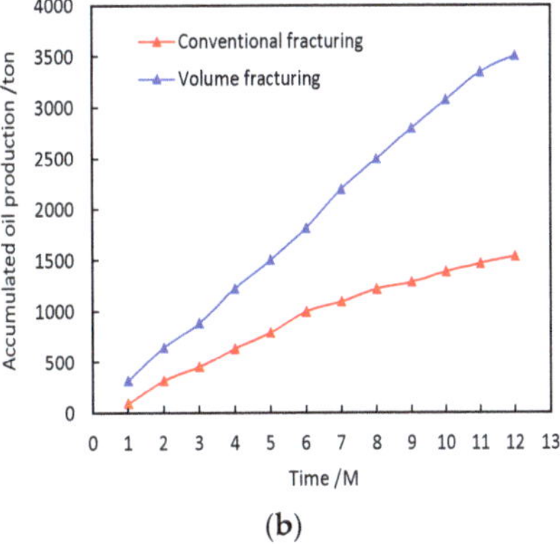

(b)

Figure 17. One-year tight oil production results in wells after volume fracturing treatment and in wells after conventional fracturing treatment. (**a**) Daily tight oil production, (**b**) accumulated tight oil production.

5. Conclusions

In this work, the performance of the volume fracturing treatment to enhance tight oil recovery in the Chang 7 tight oil reservoir, the Changqing oilfield, was studied through experimental investigation and field-scale practice. Conclusions drawn from this study are as follows:

(1) The Chang 7 tight oil reservoir in the Changqing oilfield possesses a wide distribution of hydrocarbon source rocks that are rich in organic matter. The tight oil generation intensity of the Chang 7 reservoir is estimated to be 495×10^4 km^2, and the total amount of generated hydrocarbons can be as high as 2473.08×10^8 ton. The favorable hydrocarbon source rocks in the Chang 7 reservoir ensures a high capacity of tight oil generation and accumulation.

(2) Core analysis from the pilot areas of the Chang 7 tight oil reservoir indicates poor physical properties of the reservoir formations. The average porosity of the formations is around 8–9%, and average permeability is about 0.11–0.14 mD. Due to the low porosity, low permeability, and poor pore connectivity, it is difficult to recover the tight oil from formations without fracturing treatment.

(3) Fractures developed in tight cores can significantly improve fluid flow conductivity. The increased flowability is crucial to oil recovery improvement in tight oil reservoirs. However, fractures with permeabilities too high above those of the matrix may lead to a severe water channeling problem in the formation. The injected displacing water mainly flows through the fractures with overly high permeability, leaving the tight oil in the matrix almost un-swept. Therefore, when applying fracturing treatment to improve the oil recovery in tight oil reservoirs, developing fractures with appropriate permeabilities, which increases the flowability and decreases the water channeling possibility, is important for field operators.

(4) The imbibition effect plays an important role in tight oil recovery enhancement in low-permeability porous media. Core experiments show that developed fractures can remarkably increase the imbibition rate in tight cores, as the contact area between the fracture and the matrix is significantly increased by the formed fractures. In some fractured Chang 7 tight cores, the movable oil recovery improved by the water imbibition effect can be as high as 51.32%. The imbibition effect in tight cores increases with the decrease in core permeability.

(5) A field-scale application of the volume fracturing treatment exhibits a prominent increase in tight oil production in the Chang 7 reservoir. Compared with the conventional fracturing treatment, volume fracturing can produce a larger number of fractures with more favorable properties to improve the flow conductivity. Moreover, volume fracturing treatment in horizontal wells resulted in a much larger stimulated reservoir volume around the treated wells, which can be 4.9 times greater than that of conventional volume fracturing.

Author Contributions: Conceptualization, J.W. and L.Y.; methodology, C.W.; validation, J.W., L.Y. and C.W.; formal analysis, L.Y.; investigation, L.Y.; resources, C.W. and D.C.; data curation, C.W. and D.C.; writing—original draft preparation, L.Y.; writing—review and editing, J.W.; supervision, J.W.; project administration, J.W. and C.W.; funding acquisition, J.W.

Funding: This research was funded by National Natural Science Foundation of China (51804283), the Fundamental Research Funds for the Central Universities (CUG170652) and the National Natural Science Foundation of Hubei Province (2017CFB345).

Acknowledgments: The authors would like to acknowledge the China University of Geosciences for supporting this research. We also acknowledge the National Natural Science Foundation of China (51804283), the Fundamental Research Funds for the Central Universities (CUG170652) and the National Natural Science Foundation of Hubei Province (2017CFB345) for providing research funding.

Conflicts of Interest: The authors declare no conflict of interest.

References

1. Wei, M.Q.; Duan, Y.G.; Dong, M.Z.; Fang, Q.T.; Dejam, M. Transient production decline behavior analysis for a multi-fractured horizontal well with discrete fracture networks in shale gas reservoirs. *J. Porous Media* **2019**, *22*, 343–361. [CrossRef]
2. Vengosh, A.; Jackson, R.B.; Warner, N.; Darrah, T.H.; Kondash, A. A critical review of the risks to water resources from unconventional shale gas development and hydraulic fracturing in the United States. *Environ. Sci. Technol.* **2014**, *48*, 8334–8348. [CrossRef] [PubMed]
3. Wang, Q.; Chen, X.; Jha, A.N.; Rogers, H. Natural gas from shale formation—The evolution, evidences and challenges of shale gas revolution in United States. *Renew. Sustain. Energy Rev.* **2014**, *30*, 1–28. [CrossRef]
4. Dong, D.Z.; Wang, Y.M.; Li, X.J.; Zou, C.N.; Guan, Q.Z.; Zhang, C.C.; Huang, J.L.; Wang, S.F.; Wang, H.Y.; Liu, H.L.; et al. Breakthrough and prospect of shale gas exploration and development in China. *Nat. Gas Ind. B* **2016**, *3*, 12–26. (In Chinese) [CrossRef]
5. Wang, J.J.; Wang, B.E.; Li, Y.J.; Yang, Z.H.; Gong, H.J.; Dong, M.Z. Measurement of Dynamic Adsorption-Diffusion Process of Methane in Shale. *Fuel* **2016**, *172*, 37–48. [CrossRef]
6. Wang, J.J.; Dong, M.Z.; Yang, Z.H.; Gong, H.J.; Li, Y.J. Investigation of methane desorption and its effect on the gas production process from shale: Experimental and mathematical study. *Energy Fuel.* **2017**, *31*, 205–216. [CrossRef]

7. Zhao, X.G.; Kang, J.L.; Lan, B. Focus on the development of shale gas in China—Based on SWOT analysis. *Renew. Sustain. Energy Rev.* **2013**, *21*, 603–613. [CrossRef]
8. Zhao, Y.L.; Zhang, L.H.; Luo, J.X.; Zhang, B.N. Performance of fractured horizontal well with stimulated reservoir volume in unconventional gas reservoir. *J. Hydrol.* **2014**, *512*, 447–456. [CrossRef]
9. Mao, R.; Feng, Z.; Liu, Z.; Zhao, Y. Laboratory hydraulic fracturing test on large-scale pre-cracked granite specimens. *J. Nat. Gas Sci. Eng.* **2017**, *44*, 278–286. [CrossRef]
10. Bohloli, B.; De Pater, C.J. Experimental study on hydraulic fracturing of soft rocks: Influence of fluid rheology and confining stress. *J. Nat. Gas Sci. Eng.* **2006**, *53*, 1–12. [CrossRef]
11. Vidic, R.D.; Brantley, S.L.; Vandenbossche, J.M.; Yoxtheimer, D.; Abad, J.D. Impact of shale gas development on regional water quality. *Science* **2013**, *340*, 1235009. [CrossRef] [PubMed]
12. Su, Y.L.; Zhang, Q.; Wang, W.D.; Sheng, G.L. Performance analysis of a composite dual-porosity model in multi-scale fractured shale reservoir. *J. Nat. Gas Sci. Eng.* **2015**, *26*, 1107–1118. [CrossRef]
13. Su, Y.L.; Sheng, G.L.; Wang, W.D.; Zhang, Q.; Lu, M.J.; Ren, L. A mixed-fractal flow model for stimulated fractured vertical wells in tight oil reservoirs. *Fractals* **2016**, *24*, 1650006. [CrossRef]
14. Rahm, D. Regulating hydraulic fracturing in shale gas plays: The case of Texas. *Energy Policy* **2011**, *39*, 2974–2981. [CrossRef]
15. Hummel, N.; Shapiro, S.A. Nonlinear diffusion-based interpretation of induced microseismicity: A Barnett Shale hydraulic fracturing case study nonlinear diffusion and fracturing of shales. *Geophysics* **2013**, *78*, B211–B226. [CrossRef]
16. Blanton, T.L. An experimental study of interaction between hydraulically induced and pre-existing fractures. In Proceedings of the SPE Unconventional Gas Recovery Symposium, Pittsburgh, PA, USA, 16–18 May 1982; Society of Petroleum Engineers: Pittsburgh, PA, USA, 1982. [CrossRef]
17. Maity, D.; Ciezobka, J. An interpretation of proppant transport within the stimulated rock volume at the hydraulic-fracturing test site in the Permian Basin. *SPE Reserv. Eval. Eng.* **2019**, *22*, 477–491. [CrossRef]
18. Cipolla, C.L.; Berntsen, B.A.; Moos, H.; Ginty, W.R.; Jensen, L. Case Study of Hydraulic Fracture Completions in Horizontal Wells, South Arne Field Danish North Sea. In Proceedings of the SPE Asia Pacific Oil and Gas Conference and Exhibition, Brisbane, Australia, 16–18 October 2000; Society of Petroleum Engineers: Brisbane, Australia, 2000. [CrossRef]
19. Luo, S.; Neal, L.; Arulampalam, P.; Ciosek, J.M. Flow regime analysis of multi-stage hydraulically-fractured horizontal wells with reciprocal rate derivative function: Bakken case study. In Proceedings of the Canadian Unconventional Resources and International Petroleum Conference, Calgary, AB, Canada, 19–21 October 2010; Society of Petroleum Engineers: Calgary, AB, Canada, 2010. [CrossRef]
20. Li, H.T.; Deng, J.G.; Liu, W.; Li, Y.; Lin, S. Research on Casing Deformation Failure Mechanism during Volume Fracturing for Tight Oil Reservoir of Horizontal Wells. In Proceedings of the 51st US Rock Mechanics/Geomechanics Symposium, San Francisco, CA, USA, 25–28 June 2017; American Rock Mechanics Association: San Francisco, CA, USA, 2017.
21. Rickman, R.; Mullen, M.J.; Petre, J.E.; Grieser, W.V.; Kundert, D. A Practical Use of Shale Petrophysics for Stimulation Design Optimization: All Shale Plays Are Not Clones of the Barnett Shale. In Proceedings of the SPE Annual Technical Conference and Exhibition, Denver, CO, USA, 21–24 September 2008; Society of Petroleum Engineers: Denver, CO, USA, 2008. [CrossRef]
22. Li, X.W.; Zhang, K.S.; Fan, F.L.; Tang, M.R.; Pu, X.Q.; Wang, W.X.; Li, X.P. Study and experiment on volume fracturing in low pressure tight formation of Ordos Basin. *J. Oil Gas Technol.* **2013**, *35*, 142–146, 152, 169. (In Chinese) [CrossRef]
23. Wu, Q.; Xu, Y.; Wang, T.F.; Wang, X.Q. The revolution of reservoir stimulation: An introduction of volume fracturing. *Nat. Gas Ind.* **2011**, *31*, 7–12. (In Chinese) [CrossRef]
24. Tabatabaei, M.; Mack, D.J.; Daniels, N.R. Evaluating the Performance of Hydraulically Fractured Horizontal Wells in the Bakken Shale Play. In Proceedings of the SPE Rocky Mountain Petroleum Technology Conference, Denver, CO, USA, 14–16 April 2009; Society of Petroleum Engineers: Denver, CO, USA, 2009. [CrossRef]
25. Wu, Q.; Xu, Y.; Wang, X.Q.; Wang, T.F.; Zhang, S.L. Volume fracturing technology of unconventional reservoirs: Connotation, design optimization and implementation. *Petrol. Explor. Dev.* **2012**, *39*, 377–384. [CrossRef]

26. Wu, Q.; Xu, Y.; Liu, Y.Z.; Ding, Y.H.; Wang, X.Q.; Wang, T.F. The current situation of stimulated reservoir volume for shale in US and its inspiration to China. *Oil Drill. Prod. Technol.* **2011**, *33*, 45–51. (In Chinese) [CrossRef]
27. Mayerhofer, M.J.; Lolon, E.; Warpinski, N.R.; Cipolla, C.L.; Walser, D.W.; Rightmire, C.M. What is Stimulated Rock Volume? In Proceedings of the SPE Shale Gas Production Conference, Fort Worth, TX, USA, 16–18 November 2008; Society of Petroleum Engineers: Fort Worth, TX, USA, 2008. [CrossRef]
28. Cipolla, C.L.; Warpinski, N.R.; Mayerhofer, M.J.; Lolon, E.; Vincent, M.C. The Relationship between Fracture Complexity, Reservoir Properties, and Fracture Treatment Design. In Proceedings of the SPE Annual Technical Conference and Exhibition, Denver, CO, USA, 21–24 September 2008; Society of Petroleum Engineers: Denver, CO, USA, 2008. [CrossRef]
29. Chipperfield, S.T.; Wong, J.R.; Warner, D.S.; Cipolla, C.L.; Mayerhofer, M.J.; Lolon, E.P.; Warpinski, N.R. Shear Dilation Diagnostics: A New Approach for Evaluating Tight Gas Stimulation Treatments. In Proceedings of the SPE Hydraulic Fracturing Technology Conference, College Station, TX, USA, 29–31 January 2007; Society of Petroleum Engineers: College Station, TX, USA, 2007.
30. East, L.; Soliman, M.Y.; Augustine, J.R. Methods for enhancing far-field complexity in fracturing operations. *SPE Prod. Oper.* **2011**, *26*, 291–303. [CrossRef]
31. Chen, Y.J.; Zhang, M.J.; Li, W.; Wang, H.J.; Zhang, Y. A comparative analysis of investment and benefit between conventional fracturing and fracturing by stimulated reservoir volume (SRV): Case history of gas/shale gas wells in the southern Sichuan basin. *Nat. Gas Ind.* **2014**, *34*, 1–5. (In Chinese) [CrossRef]
32. Yu, B.M.; Xu, P.; Chen, J. Study on the seepage characteristics in fractured network porous media. *J. Xi'an Shiyou Uni. (Nat. Sci. Ed.)* **2007**, *22*, 21–22, 25, 174–175.
33. Lei, Q.; Xu, Y.; Jiang, T.X.; Ding, Y.H.; Wang, X.Q.; Lu, H.B. "Fracture network" fracturing technique for improving post-fracturing performance of low and ultra-low permeability reservoirs. *Acta Pet. Sin.* **2009**, *30*, 237–241. [CrossRef]
34. Niu, X.B.; Feng, S.B.; You, Y.; Liang, X.W.; Xin, H.G.; Dan, W.D.; Li, T.Y.; Ren, J.S. Fracture extension and distribution pattern of volume fracturing in tight reservoir: An analysis based on actual coring data after fracturing. *Oil Gas Geol.* **2019**, *40*, 669–677. (In Chinese) [CrossRef]
35. Bai, X.H.; Qi, Y.; Lu, H.J.; Duan, P.H.; Gu, Y.L.; Wu, P.R. Optimization design for volume fracturing of horizontal wells in tight oil reservoir of Ordos Basin. *Oil Drill. Prod. Technol.* **2015**, *37*, 83–86. (In Chinese) [CrossRef]
36. Fan, J.M.; Yang, Z.Q.; Li, W.B.; Wang, C.; He, Y.A. Assessment of fracturing treatment of horizontal wells using SRV technique for Chang-7 tight oil reservoir in Ordos Basin. *J. China Univ. Pet. (Ed. Nat. Sci.)* **2015**, *39*, 103–110. (In Chinese) [CrossRef]
37. Li, Z.Q.; Zhao, J.Z.; Hu, Y.Q.; Ren, L.; Li, Y. Productivity forecast of tight oil reservoirs after multi-zone stimulated reservoir volume fracturing. *Pet. Geol. Recovery Effic.* **2016**, *23*, 134–138. (In Chinese)
38. Li, Z.X.; Li, J.; Qu, X.F.; Li, X.W.; Lei, Q.H.; Fan, J.M. The experiment and recognition of the development of Chang 7 tight oil in Ordos Basin. *Nat. Gas Geosci.* **2015**, *26*, 1932–1940. (In Chinese) [CrossRef]
39. Mirzaei-Paiaman, A.; Kord, S.; Hamidpour, E.; Mohammadzadeh, O. Scaling one-and multi-dimensional co-current spontaneous imbibition processes in fractured reservoirs. *Fuel* **2017**, *196*, 458–472. [CrossRef]
40. Ghanbari, E.; Dehghanpour, H. The fate of fracturing water: A field and simulation study. *Fuel* **2016**, *163*, 282–294. [CrossRef]
41. Xiao, L.; Xiao, Z.X. Analysis of methods for determining NMR T2cutoff and its applicability. *Prog. Geophys.* **2008**, *23*, 167–172. (In Chinese) [CrossRef]
42. Shao, W.Z.; Ding, Y.J.; Xiao, F.; Mu, H.W.; Liu, H.B.; Cai, H.; Dong, X.F. On the method of determining T2 cutoff value with the T2 spectrum characteristics. *Well Logging Technol.* **2009**, *33*, 430–435. (In Chinese)
43. Li, Z.C. Classification of fractured reservoirs by T2 spectrum nuclear magnetic resonance log. *Geophys. Prospect. Pet.* **2001**, *40*, 113–118. (In Chinese)

© 2019 by the authors. Licensee MDPI, Basel, Switzerland. This article is an open access article distributed under the terms and conditions of the Creative Commons Attribution (CC BY) license (http://creativecommons.org/licenses/by/4.0/).

Article

Mechanisms and Influence of Casing Shear Deformation near the Casing Shoe, Based on MFC Surveys during Multistage Fracturing in Shale Gas Wells in Canada

Yan Xi [1,*], Jun Li [1,*], Gonghui Liu [1,2], Jianping Li [3] and Jiwei Jiang [1]

1 College of Petroleum Engineering, China University of Petroleum-Beijing, Beijing 102249, China; lgh1029@163.com (G.L.); jiangjiwei0603@126.com (J.J.)
2 College of Mechanical Engineering and Applied Electronics Technology, Beijing University of Technology, Beijing 100022, China
3 CNPC Logging Co., LTD, Xi'an 710000, China; cpl-ljp@cnpc.com.cn
* Correspondence: garfield0510@163.com (Y.X.); lijun446@vip.163.com (J.L.); Tel.: +86 010-89731225 (Y.X.); +86 010-89731225 (J.L.)

Received: 12 December 2018; Accepted: 24 January 2019; Published: 24 January 2019

Abstract: Casing shear deformation has become a serious problem in the development of shale gas fields, which is believed to be related to fault slipping caused by multistage fracturing, and the evaluation of the reduction of a casing's inner diameter is key. Although many fault slipping models have been published, most of them have not taken the fluid-solid-heat coupling effect into account, and none of the models could be used to calculate the reduction of a casing's inner diameter. In this paper, a new 3D finite element model was developed to simulate the progress of fault slipping, taking the fluid-solid-heat coupling effect during fracturing into account. For the purpose of increasing calculation accuracy, the elastoplastic constitutive relations of materials were considered, and the solid-shell elements technique was used. The reduction of the casing's inner diameter along the axis was calculated and the calculation results were compared with the measurement results of multi-finger caliper (MFC) surveys. A sensitivity analysis was conducted, and the influences of slip distance, casing internal pressure, thickness of production and intermediate casing, and the mechanical parameters of cement sheath on the reduction of a casing's inner diameter in the deformed segment were analyzed. The numerical analysis results showed that decreasing the slip distance, maintaining high pressure, decreasing the Poisson ratio of cement sheath, and increasing casing thickness were beneficial to protect the integrity of the casing. The numerical simulation results were verified by comparison to the shape of MFC measurement results, and had an accuracy up to 90.17%. Results from this study are expected to provide a better understanding of casing shear deformation, and a prediction method of a casing's inner diameter after fault slipping in multistage fracturing wells.

Keywords: multistage fracturing; shear deformation; numerical simulation; fluid-solid-heat coupling; Multi Finger Caliper

1. Introduction

Horizontal well and multistage fracturing are the two key techniques for shale gas development, by creating complex fracture networks within tight formations [1–5]. Due to the fact that tens of thousands of cubic meters of fracturing fluid are pumped into the downholes of shale gas wells and injected into the matrix of shale reservoirs, geostatic stress can be changed due to the elastic response of the rock mass to hydraulic fracturing. Pore pressure can also be changed due to fluid diffusion along a permeable fault zone [6]. As a result of this, the casing string is exposed to a complex mechanical environment in the downhole, and therefore the risk of casing deformation increases dramatically [7–9]. Previous studies have shown that casing deformations were observed during multistage fracturing in the United States and China, where the commercial development of shale gas is underway and investment is on the rise [10,11]. Casing deformation has created a lot of problems for shale gas well completion and development, for example, bridge plugs could not be run to the projected depth, and normal stimulation operations were unable to be carried out. As a result of this, some fracturing sections either needed to be repaired, which increased the cost of well completion, or could only be abandoned, which decreased the productivity of the shale gas wells. In addition, based on previous studies about casing deformation in conventional oil and gas wells, it is known that casing deformation always intensifies over time, which could lead to security issues for the public or the shut-in of the well. Therefore, there is an urgent need to analyze the mechanisms and propose effective solutions for casing deformation.

Multi-finger calipers (MFC) and lead molds are two effective tools for monitoring the deformation features of the deformed part of a casing [12,13]. Based on investigations into casing deformation by using both tools in China, there were four different types of casing deformations that were delineated, including extrusion deformation, shear deformation, bending deformation, and buckling. Casing shear deformation represented the largest portion of all of the deformed points. Statistic data showed that: (a) up to March 2016, a total of 90 horizontal wells were successfully fractured in Weiyuan-Changning block, Sichuan basin, casing deformation occurred in 32 wells, and 47 deformed points were found [14,15]. 61.7% of all the casing deformed points were examplse of shear deformation [16], and (b) by the end of May 2018, six horizontal wells were successfully fractured in Weirong block, where casing deformation occurred in five wells and 17 deformed points were found. Most of the deformed points were due to shear deformation. Serious casing deformation has occurred during multistage fracturing in shale gas wells in Simonette, Canada, although there is no previous public data. MFC surveys were conducted in five pads, including 28 wells in this study, and the statistical data showed that 52.2% of all the deformed points were due to shear deformation. From the above data, it can be seen that the study of the mechanisms of casing shear deformation appears to be particularly important to the research on casing deformation during multistage fracturing in shale gas wells.

Many related studies have been carried out, but most of them emphasized the inducement of casing shear deformation [17]. Based on the previous research work, faults were easily reactivated when fracturing fluid flew into the cracks in the formation, and were more likely to slip along the unstable bedding planes or natural fractures under the action of their own gravity or external forces [18,19]. Then, the casing strings which passed through the faults were sheared. Qian et al. [20] pointed out that formation stress changed, due to opened and propped hydraulic fractures, and caused natural fractures to open or slip, which increased the risk of casing shear deformation. According to the microseismic data, Zoback and Snee [21] believed that the high pore pressure generated during hydraulic fracturing operations induced slip on preexisting fractures and faults with a wide range of orientations. Meyer et al. [11] analyzed the seismic data and multi-finger caliper data, and suggested that shear failure of pre-existing faults was likely the main cause of casing deformation. Some scholars have proposed a similar viewpoint and complimented the study, pointing out that when the borehole trajectory was inclined upward along the formation, after fracturing fluid flowed into the shale beddings, the fault could slip because of the effect of gravity [15,16]. The mechanisms of fault slippage during or after multistage fracturing were discussed in most of the current research, but few have calculated the variation of the casing's inner diameter, which was the determining factor of whether the bridge plug could pass through the deformed part of the casing, and therefore should be the evaluation basis for effective solutions to address the problem of casing shear deformation.

Analysts can more precisely identify the deformation situation of a casing's deformed parts by using an MFC tool. Some scholars have expounded upon the application of an MFC tool in conventional oil and gas wells, which indicated that MFC data could accurately reflect the actual deformation of the casing string in the well [12,13,22]. Despite the clear benefits of MFC surveys, this kind of technique remains challenging to implement in extensive regional oil and gas fields, due to the significant cost of a full system. As a result, there has been little research providing the evidence of MFC data, and only some conducted prospective studies. Marc [22] collected MFC measurement results from 30 wells from 2003 to 2013, showing that the shear deformation features were localized over a relatively short length (several feet), and resulted from a relative displacement of the upper part of the well compared to the lower part. According to the measurement results of the MFC tool, some mathematical and numerical models were established to simulate the progress of fault slipping, so as to calculate the degree of casing shear deformation. Gao et al. [23] analyzed the characteristics of casing shear deformation, established a 3D finite element model to simulate the stress-strain status and the deformation process, and pointed out that no cementing near the position of slipping was beneficial to the mitigation of casing shear deformation. Chen et al. [16] presented a mathematical model for establishing the relationship between microseismic moment magnitude and degree of casing sheared deformation. Guo et al. [24] developed a numerical model and calculated the influences of slip distance, slip angle, and the mechanical parameters of cement sheath on casing stress. But few of the above studies analyzed the difference between measurement results and actual deformation, and have not put forward solutions to casing shear deformation and showed their validity in engineering in practice. Therefore, it is still a great challenge to solve the problem of casing shear deformation.

In this study, MFC surveys on casing deformation in Canada were implemented, and the mechanisms of casing shear deformation that occurred at the interface between different layers were studied. A new 3D numerical model was developed to simulate the progress of fault slipping, the variation of a casing's inner diameter along the axis was calculated based on the analysis of MFC surveys. The numerical simulation results were verified through the measured data. Six influential factors, including the slip distance, casing inner pressure, thickness of production casing and intermediate casing, and mechanical parameters of cement sheath were analyzed. Furthermore, proposals for mitigation or avoidance of casing shear deformation were suggested, and some of them were applied to the engineering in practice and were proven to be effective.

2. Overview of Casing Shear Deformation in Simonette

2.1. Field Description

Simonette is one of the most important shale gas blocks in Canada, and is located in the west of Alberta Basin. The shale reservoir is situated in the Upper Devonian Duvernay Formation (Woodbend Group), which is composed of multicyclic units of black organic-rich shale and bituminous carbonates, ranging between 25 and 60 m in thickness [25], and extends throughout most of central Alberta. The vertical depth of the reservoir is approximately 3800–3950 m, and the Duvernay Formation exhibits a westward increase in thermal maturity from the immature to the gas-generation zone [26]. Much of the Duvernay Formation in the area of optimal maturity is overpressured, with noted cases beyond gradients of 1.9 MPa/100m in deep basin settings [25,27]. And the temperature gradient is approximately 3.3–3.7 °C/100 m.

Figure 1 displays the typical well architecture deployed for the development of Simonette. The main drilling phases are: (1) a 349 mm section drilled from the wellhead to approximately 620 m; (2) a 222 mm section drilled from the previous shoe down to the Ireton formation (the true vertical depth (TVD) is approximately 3750 m), including the vertical section and part of indication section; (3) a 171 mm section drilled from the previous shoe down to the Duvernay formation (the TVD is approximately 3885 m), including part of indication section and the whole horizontal section. The depth of the interface between the Nisku and Ireton formations is about 3742 m. It is worth noting that this interface happens to cross-cut the intermediate casing, which is close to the casing shoe.

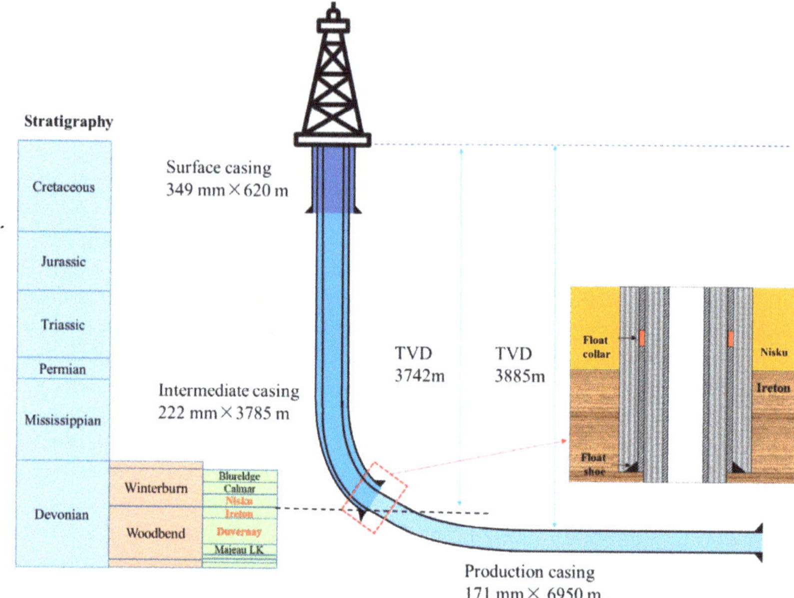

Figure 1. Geological stratification and well structure.

2.2. Casing Shear Deformation

Casing perforation completion and bridge plug staged fracturing were adopted in shale gas wells in Simonette. The number of fracturing segments was approximately 40, each stage was injected with about 1500 m^3 of fracturing fluid, with a displacement of 12–14 m^3/min and pumping pressure of over 70 MPa. 5 pads including 28 wells were investigated by MFC surveys. Casing deformation occurred in 16 wells during multistage fracturing. 23 deformed points were found, and there were five different types of deformed points, including extrusion deformation, shear deformation, bending deformation, buckling, and casing holes, as shown in Figure 2. Statistical data showed that 52.2% of all of the deformed points were shear deformation.

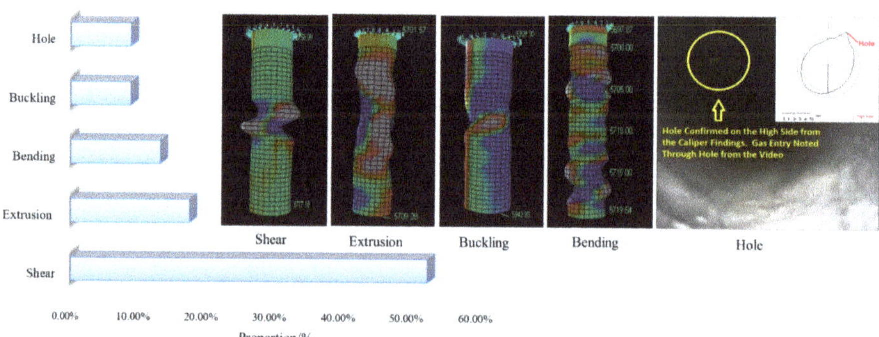

Figure 2. Five types of casing deformation and their respective proportions.

Furthermore, the shear deformed points can be classified into two types according to the positions of occurrence: (a) the first type of shear deformed points was located at the position of the interface between the Nisku and Ireton formations, and accounted for 75% of the total shear deformed points;

(b) the second type of shear deformed points was located in the horizontal segment, and accounted for 25%. Therefore, it is very meaningful to clarify the mechanism of the first type of casing shear deformation, which is also the aim of this study.

From the introduction above, it is known that the production and intermediate casings were cross-cut by the interface between the Ireton and Nisku formations. According to the statistical data about the depth of the first type of casing shear deformation, all of the shear deformed points occurred at the interface and in the bottom production casing above the casing shoe, as shown in Figure 3. In addition, the lengths of the deformed parts were relatively short, about 1.2–2 m.

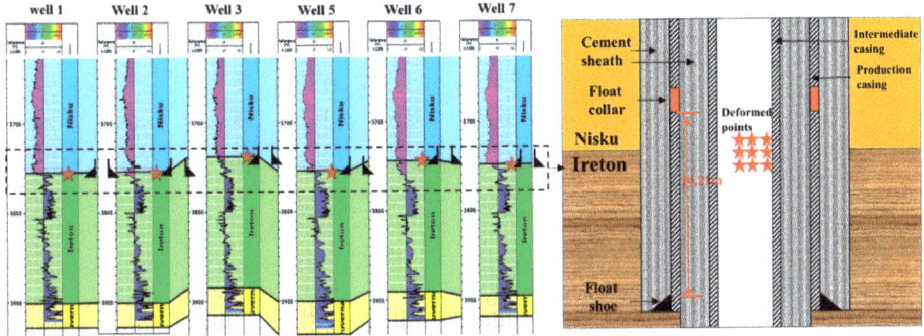

Figure 3. Locations of the first type of casing shear deformed points.

2.3. Difference Between Measurement Results and Actual Shear Deformation

MFC data can be used to assess the inner conditions of the casing after deformation. As a result of this, shear deformation features identified in the casing turned out to be critical. During the research process, most scholars considered an MFC tool that was centered in the casing, and ignored the transition from measurement results to actual deformation morphology. Indeed, when the MFC tool string runs through the deformation position in the casing, the two centralizers force the middle of the tool string (including the caliper tool) off-center (Figure 4a), which affects the measurement data, and therefore the 3D morphology reflected by the rough data is incorrect.

The difference between the shapes reflected from rough measurement results and actual deformation is illustrated in Figure 4b. From the 3D views of the rough data, it can be seen that the casing appears to be sheared in the upper and lower portions, but the reality shown after data transition indicates that the casing was sheared only at one side. This illustrates that negligence of this condition can lead to wholly misleading results. The deviation of the casing was defined as the degree of casing shear deformation, and statistical data showed the degree had already reached almost 45 mm. The deformation degree (Figure 4b) can be used to describe the slip distance to some extent, but can not be used to measure the reduction of the casing's inner diameter. Therefore, the deformation degree can not be used as the basis of evaluating whether the bridge plug could pass through the deformed part. As a result of this, the relationship between slip distance and the reduction of the casing's inner diameter should be established.

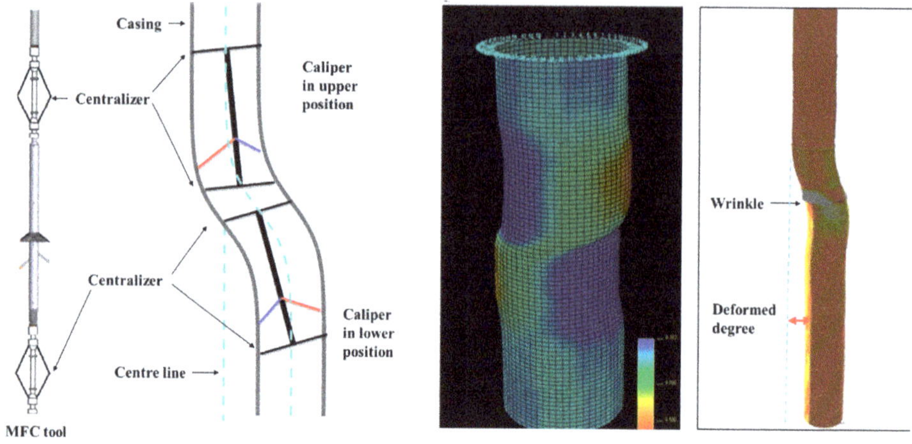

(**a**) Schematic diagram (**b**) Measurement results and actual conditions

Figure 4. Schematic diagram and comparison of measurement results and actual conditions.

3. Mechanisms of Casing Shear Deformation induced by Multistage Fracturing

3.1. Mechanisms of Casing Shear Deformation

Previous studies have proven that fault slipping was the main reason for casing shear deformation. On the basis of worldwide observations, casing shear deformation caused by fault slipping can be classified into three main categories:

(1) Shear deformed points overlap with natural fractures (type 1, Figure 5). This is the most common mechanism of casing shear deformation, and was identified in almost all of the shale gas fields [14–17,21]. This type of casing shear deformation is not necessarily restricted to the high shear stress area, and the position of the shear deformed points are more directly related to the location and orientation of natural fractures, rather than the location of the crustal stress.

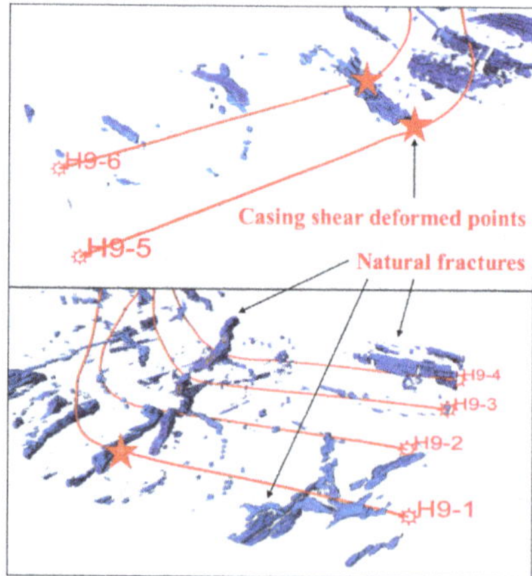

Figure 5. Deformed points overlap with natural fractures.

(2) Shear deformed points appeared near the landing point (type 2, Figure 6). The landing point is the starting point of the horizontal segment. Nearly all the trajectories of the wells which incurred

casing deformation of this type were inclined upward along the formation. Some scholars [14,21] have believed that during the progress of fault slipping, the gravity of the fault plays an important role in activating the faults after the fracturing fluid flew into the interface of bedding planes.

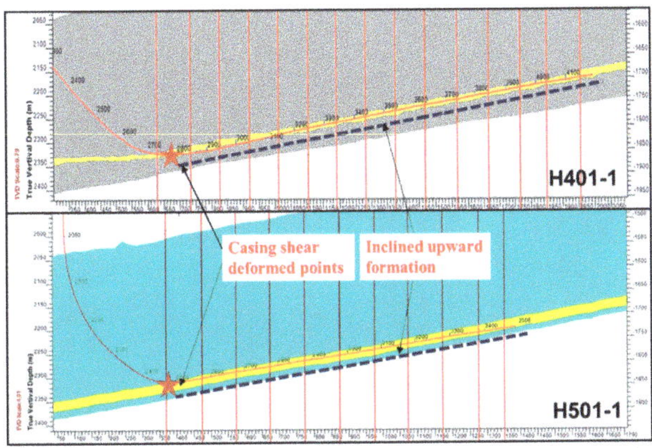

Figure 6. Deformed points near the landing point.

(3) Shear deformed points at the interface between different layers (type 3). Shear deformed points located at the position of the interface between the Nisku and Ireton formations are of this type. The interface between different formations is opened, due to the high bottom hole pressure during the operation of well cementation, and the friction coefficient between the formations decreases dramatically. During the progress of multistage fracturing, the fault below the interface is easily activated and it can slip along the interface.

3.2. Verification of Fault Slipping

Suppose that the interface between Nisku and Ireton formations was a weak interface, especially after the interface was opened and the cement slurry flowed into it. The shear and normal stress on the oriented plane can be calculated by coordinate transformation. The formation/fault would slip when

$$\tau > f_n \sigma_n + S, \tag{1}$$

where τ represents the shear stress applied to the fault (MPa); f_n represents the coefficient of friction, dimensionless. And according to Zoback [28], the value range of f_n is from 0.6 to 1. σ_n represents the effective normal stress (MPa); S represents the rock cohesive strength (MPa).

Based on Biot's law [29], σ_n can be expressed as

$$\sigma_n = S_n - P_p, \tag{2}$$

where P_p is the pore pressure in MPa, and S_n is the normal stress perpendicular to the interface.

The pore pressure increases at the bedding plane during fracturing; when it meets the formula requirements, the minimum increment of pore pressure (MPa) is [28]:

$$P' = \frac{S}{f_n} + \sigma_H + (\sigma_H - \sigma_v)\left(\sin^2\psi - \frac{\sin\psi\cos\psi}{f_n}\right), \tag{3}$$

where $\Delta\rho_w$ is equal to the yield density in (g/cm^3). ψ is the angle between the interface and the maximum horizontal principal stress (degree). The normal stress has been resolved into horizontal (σ_H) and vertical (σ_v) components.

The results calculated according to the conditions given in this study (Section 4.3) showed that the critical pore pressure was 81.2 MPa. In addition, the downhole pressure was as high as 115 MPa, indicating that the fault was likely to slip at excessive pore pressure.

In order to demonstrate this further, a microseismic technique was used to monitor the fault slipping which occurred at the position of the interface between the Nisku and Ireton formations [30–33]. From the depth view of 9-31 well, it can be seen that hydraulic fractures already extended to the interface (Figure 7a). Some microseismic events appeared at different locations, which were far apart from each other, but all along the interface, which indicated that the interface was opened during the process of fracturing, and the fault was likely to slip. Microseismic data of the 12-6 pad showed that events were found at the top of Ireton formation (Figure 7b), which was direct evidence of the fault slipping which occurred at the position of the interface between the Nisku and Ireton formations.

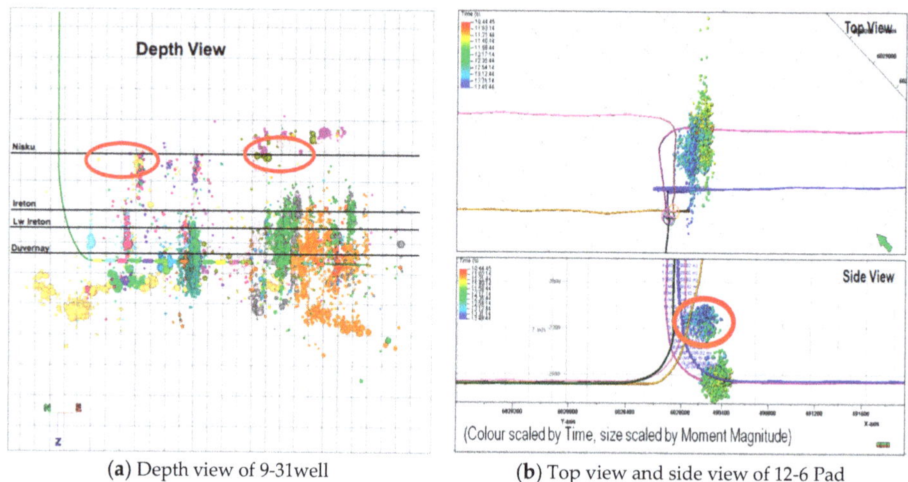

(**a**) Depth view of 9-31well (**b**) Top view and side view of 12-6 Pad

Figure 7. Microseismic data.

4. Numerical Simulation

In order to calculate the variation of the casing's inner diameter, a numerical model was developed and the influential factors were taken into account. Simulations were carried out by using the commercial software Abaqus (6.14-1), which can be used to perform and post-process simulations of various cases, and for statistical sensitivity analysis. The measurement results of the MFC tool are used to confirm the validity of the proposed Finite Element Model (FEM).

4.1. Model Geometry and Discretization

Physical model. The assembly, which contains a production casing-cement sheath-intermediate casing -cement sheath formation, located above the intermediate casing shoe, was selected as the research object. It was assumed that the casings were centered and the cement sheaths were integral. The formation includes two blocks, the upper block which represented the Nisku formation was the fixed part, and the lower block which represented Ireton formation was the slip part, as shown in Figure 8a.

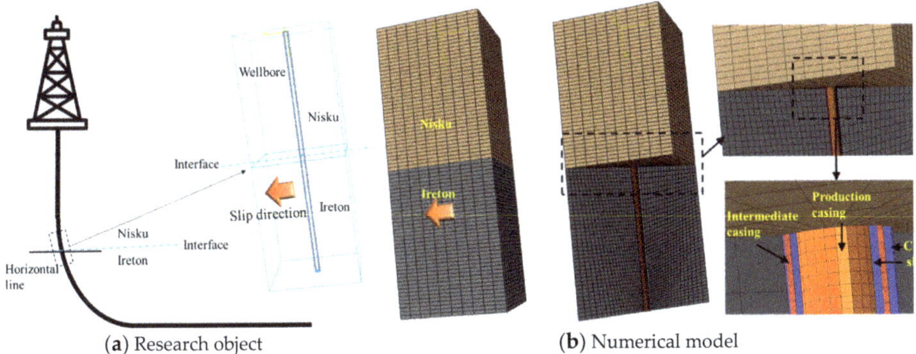

(**a**) Research object (**b**) Numerical model

Figure 8. Research object and numerical model of fault slipping.

Numerical model. A three-dimensional (3D) nonlinear FEM was established to simulate the slip progress of the fault and the mechanical behaviors of the casing shear deformation, as shown in Figure 8b. As assumed above, the casings were centered, and the cement sheaths showed complete integrity. The outer diameter of the model was 3 m × 3 m × 8 m, which was ten times greater than the size of the borehole, thus allowing the influence of the boundary to avoid the effects on the stress.

Discretization. During the simulation, the materials in the model were chosen assuming that the casing followed the elastically-perfect plastic constitutive relationship with the Von Mises yield criterion, and the cement sheath and formation followed the elastic-plastic constitutive relationship with the Mohr-Coulomb criterion. In order to reflect the casing deformation accurately, the solid element (3D stress, C3D8R) is used to analyze the cement sheath-intermediate casing-cement sheath formation, and the shell element (shell, S4R) is used to study production casing. During the progress of discretizing the finite element model, in order to increase the computational accuracy, the structured grid and variable density meshing method is applied to the model.

4.2. Boundary Conditions and Simulation Steps

Boundary Conditions. In terms of load and constraint sets, the composite boundary of the upper part was fixed by imposing displacement constraints, and the slip displacement of the Ireton part is imposed on the corresponding formation's surface. Through finite element predefined field function, far-field stress was applied, and the hydraulic pressure was added to the inner wall of the casing. The stress field of the research object, which was part of the inclination segment, could be obtained by three-time rotations from the original coordinate system, and this was then applied to the numerical software (Appendix A). The pressure of fracturing fluid was applied to the casing's inner pressure.

In order to reflect the actual conditions of the casing, the transient temperature change of the model was taken into account. The initial temperature of assembly was equal to formation temperature, and the outside surface of the model was one of the temperature boundaries and was set as a stable thermal source. The casing's inner wall was another temperature boundary, and its temperature was equal to that of the fracturing fluid, which should be calculated by a wellbore temperature field model during fracturing (Appendix B). The variation of time-dependent temperature was the input for the numerical model as a dynamic boundary.

Simulation Steps. At the first step, the temperature variations of the well are calculated when imposing the dynamic boundary at the inner wall of the production casing. Subsequently, the calculated temperature distribution, geo-stress, and the internal pressure are applied to the assembly, and the initial state of equilibrium is reached. Lastly, the lower block slipped and casing deformation occurred, then the displacement of casing's inner wall was calculated. During the third step, for sensitivity purposes, the slip distance, casing inner pressure, the production and intimate casing thickness, and mechanical parameters of cement sheath will be changed in order to find the optimal parameters during simulation.

4.3. Geological and Mechanical Parameters

As a horizontal shale gas well in Simonette, 12-10 wells were drilled to the maximum vertical depth of 3920 m, with a horizontal segment length of 2450 m. The measured and true vertical depths of the interface between the Nisku and Ireton formations were 3786 m and 3742 m, respectively. The horizontal minimum in-situ stress and vertical in-situ stress gradients were 2.3 MPa/100 m, 2.0 MPa/100 m, and 2.5 MPa/100 m, respectively. The hole inclination angle is 24°, and the well azimuth angle is 78°. The pumping pressure of fracturing was 77 MPa, and the fracturing fluid friction is approximately 10 MPa. The fracturing fluid density was 1.28 g/cm^3, then the casing's inner pressure in the interface was about 115 MPa. The discharge of fracturing fluid is 12 m^3/min, and the fracturing time is 4 h. Other geological and mechanics parameters are shown in Tables 1 and 2.

Table 1. Geometric and mechanical parameters of the assembly.

Component	Outer Diameter (mm)	Young Modulus (GPa)	Poisson's Ratio	Cohesive Strength (MPa)	Internal Friction Angle(°)
Production Casing	139.7	210	0.3	\	\
Cement Sheath (2)	171	10	0.17	8	27
Intermediate Casing	193.7	210	0.3	\	\
Cement Sheath (1)	222	10	0.17	8	27
Formation	\	22	0.23	5	39

Table 2. Thermodynamic Properties of the assembly.

Materials	Coefficient of Heat Conduction (W·(m·°C)$^{-1}$)	Specific Heat (J·(kg·°C)$^{-1}$)	Density (kg·m^{-3})	Coefficient of Thermal Expansion (10^{-6}·°C^{-1})
Casing	45	461	7800	13
Cement sheath	0.98	837	3100	11
Formation	1.59	1256	2600	10.5

5. Results

5.1. Engineering Verification of the Numerical Simulation

Figure 9 shows the distribution of casing strain after fault slipping, which indicates that the stain concentrations appear at the position of the slip interface, and shows symmetrical distribution along the center line of the casing. The overall deformed shape of the casing after fault slipping is similar to the actual deformation, as shown in Figure 4b. Line a1-a2 and line b1-b2 are perpendicular, which can represent the casing's inner diameter. Because the plane a1-a2-a'2-a'1 is parallel to the direction of fault slipping, the greatest variation of the casing's inner diameter occurred in this plane after fault slipping. As a result of this, line a1-a'1 and a2-a'2 were selected to accurately assess the variation of the inner diameter along the whole casing.

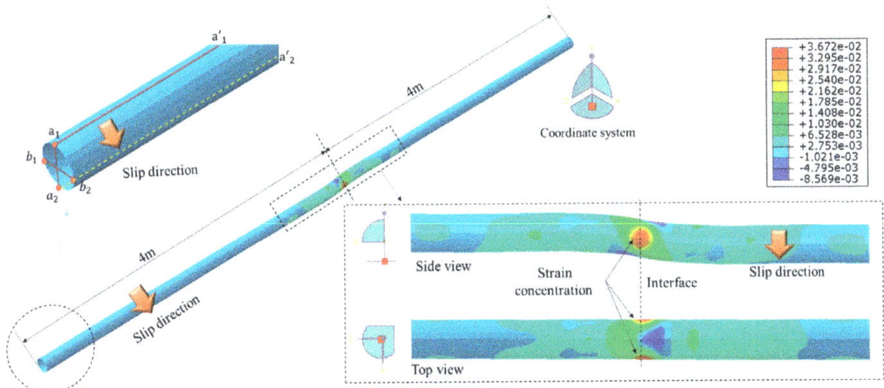

Figure 9. Distribution of casing strain (dimensionless).

Figure 10 shows the displacement curves of line a1-a'1 and a2-a'2 after fault slipping. Both curves were divided into three sections, as shown in Figure 10b–d. The displacement differences of the two curves in the first and third part are relatively small, which means that the inner diameter of the casing is almost unchanged. There is a rule for these two parts that the closer the distance from the slip interface, the more obvious the difference between the two lines. The displacement differentiation of the two lines is very obvious in the second part, which demonstrates the casing's inner diameter changes significantly, and structural distortion occurs. Therefore, the second part is the most essential part for deciding whether the bridge plug can pass the deformed part.

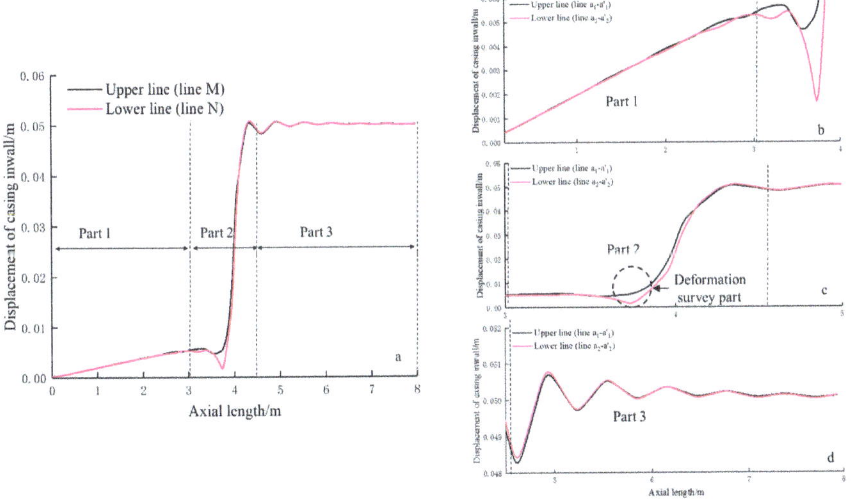

Figure 10. Variation of displacement along the casing.

The length of the second part is 1.38 m, which is in accord with the statistical rule. In order to assess the accuracy of the calculation, the simulated and measurement results at the special position are compared, as shown in Figure 11. It is shown that the result of the cross-section at the position of deformed part shows remarkably consistent findings.

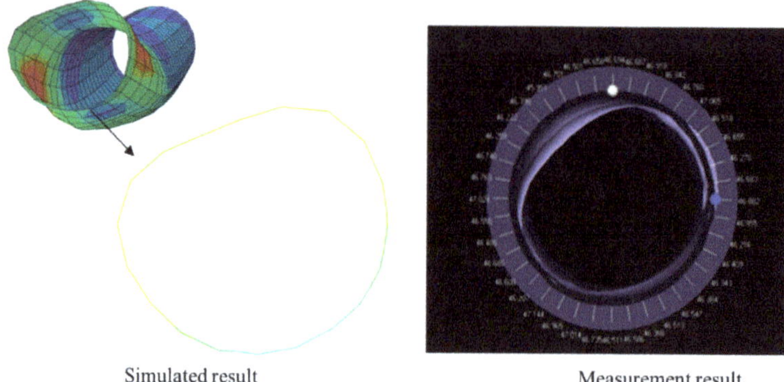

Figure 11. Comparison of the simulation and measurement results.

On the basis of the above analysis, the variation of diameter along the whole casing was calculated, as shown in Figure 12, which indicates that the most dramatic change appears at the position of the slip interface. That the maximum reduction of casing inner diameter is 4.06 mm leads to a drop of 3.35% compared to the original inner diameter.

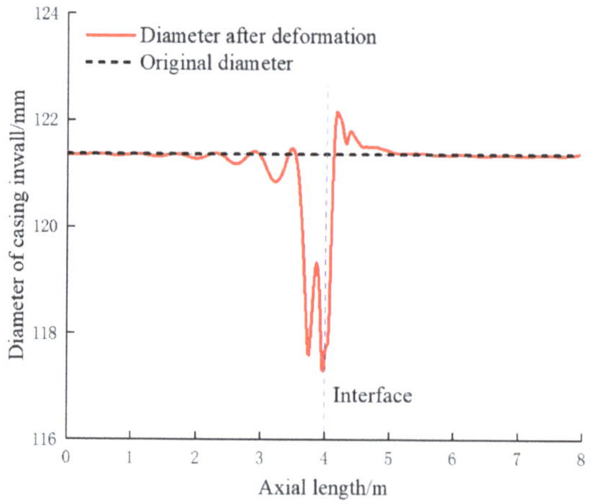

Figure 12. Variation of diameter along the casing.

5.2. Sensitivity Analysis of Casing Shear Deformation

5.2.1. Influence of Slip Distance

Previous studies have shown that with the increase of slip angle (0–90°), fault slipping had a growing influence on casing shear deformation [16]. Consider the worst-case scenario, which is that the slip angle is 90°, the influence of slip distance on the variation of casing's inner diameter was calculated, as shown in Figure 13. It can be concluded that with the increase of the slip distance, the reduction of the casing's inner diameter increases clearly, and the deformation becomes more and more complex and severe. It is worth noting that the number of concave areas in the curves changes from one to two (Figure 14), which means that the larger slip distance makes it harder for the bridge plug to pass through the deformed part. Also, it is the cause of the appearance of the wrinkle in the inner wall of casing, as shown in Figure 4b.

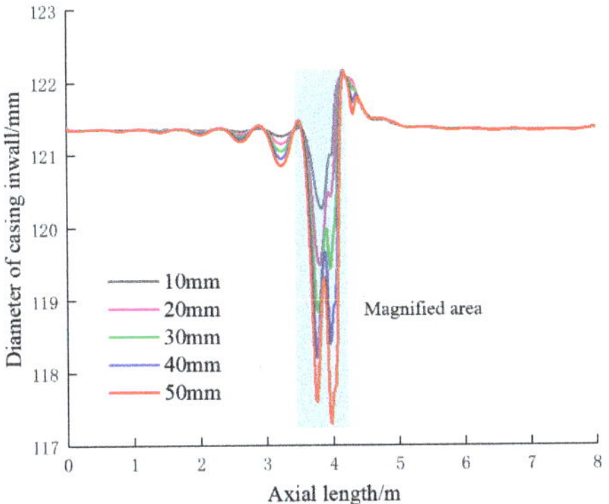

Figure 13. Variation of diameter under different slip distances.

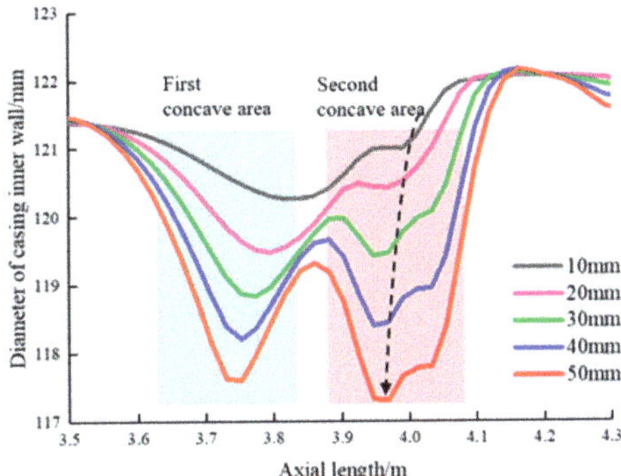

Figure 14. Magnified area.

5.2.2. Influence of Casing Inner Pressure

Fault slipping could happen during or after fracturing, as the casing's inner pressure is different in different stages. Figures 15 and 16 show the variation of diameter along the casing under different inner pressures. It can be seen that with the increase of inner pressure, the reduction of the casing's inner diameter decreases, which indicates that higher inner pressure is beneficial to maintain the casing's integrity. It should be noted that the casing's inner pressure has little impact on the first concave area in the curves, while it has a relatively significant impact on the second concave area. When the casing's inner pressure is zero, the reduction of diameter is the greatest (4.90 mm), and compared with the reduction when the casing's inner pressure is 115 MPa (the reduction is 4.06 mm), the increase of the diameter reduction is 20.7%. The reason for this is mainly because the casing's inner pressure has a dramatic impact on the equivalent stiffness of the casing string. The higher the casing's inner pressure, the higher the equivalent stiffness.

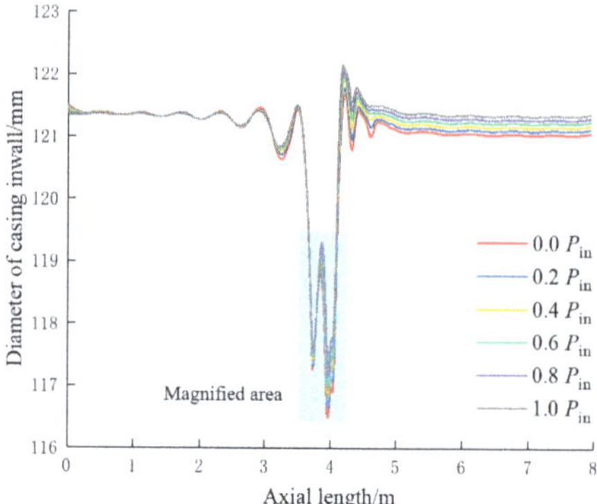

Figure 15. Variation of diameter under different casing inner pressures.

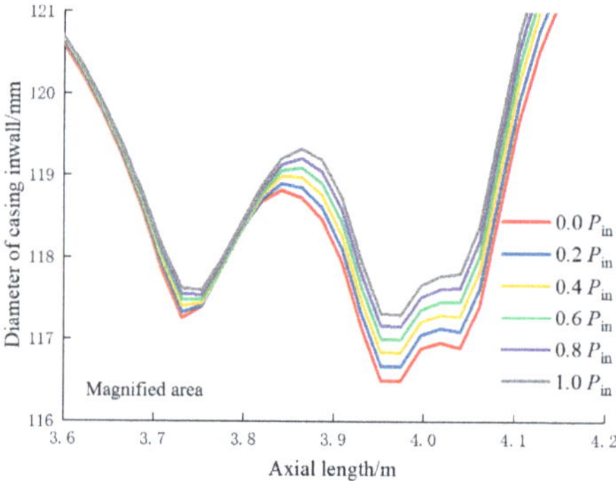

Figure 16. Magnified area.

5.2.3. Influence of Casing Thickness

With the increase of casing thickness, the shear resistance strength increases. Because the research object contains two casings, including intermediate and production casing, the thickness of both types of casings are changed, so as to evaluate the effectiveness of increasing the casing thickness. Figure 17 shows the variation of the casing's inner diameter under different production casing thicknesses. It can be seen that with the increase of the casing thickness, the reduction of casing's inner diameter is almost unchanged in the first concave area, but decreases clearly in the second concave area (Figure 18). Figure 19 shows the reduction of the casing's inner diameter with the variation of the intermediate casing thickness, and it can be seen that with the increase of the casing thickness, the reductions of the casing's inner diameter decreases nearly linear in both concave areas. Comparing the two methods, increasing the thickness of the intermediate casing above the casing shoe is the more effective.

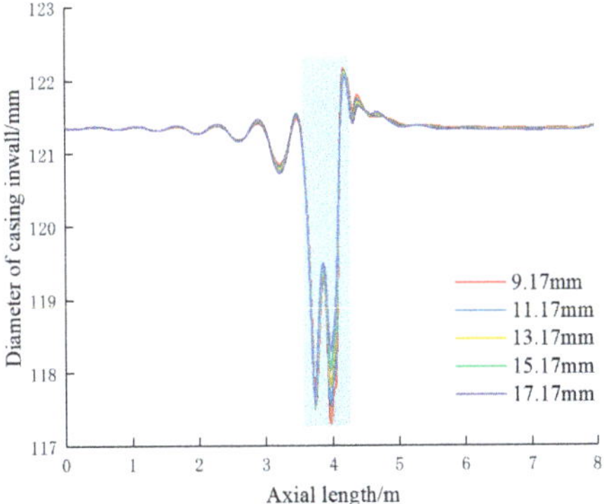

Figure 17. Variation of diameter under different production casing thickness.

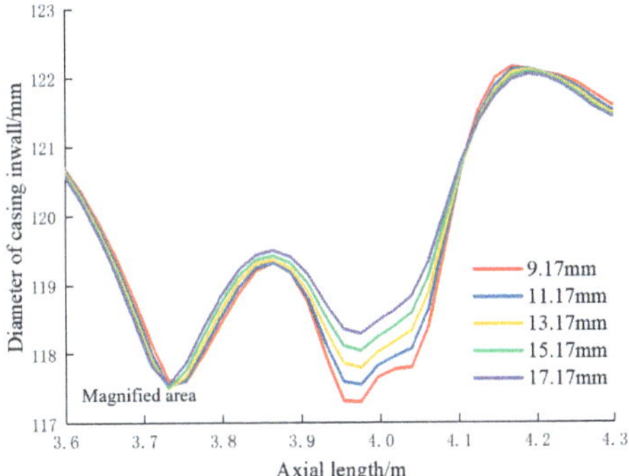

Figure 18. Magnified area.

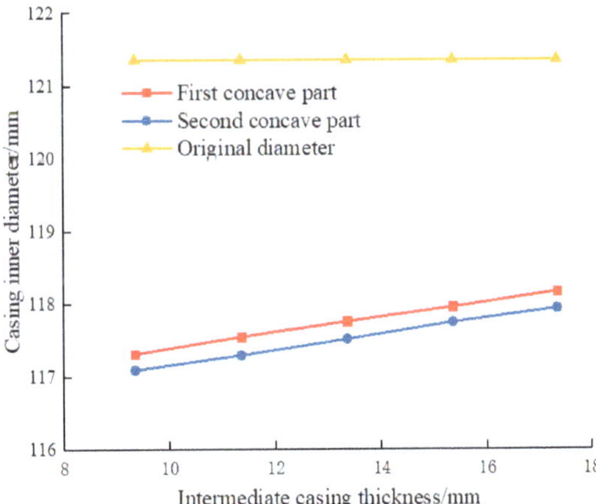

Figure 19. Variation of diameter under different intermediate casing thickness.

5.2.4. Influence of Cement Sheath Mechanical Parameters

The mechanical parameters of the cement sheath can be adjusted by using different cement slurry formulas, and have an impact on the reduction of the casing's inner diameter. Figure 20 shows the variation of diameter along the casing under different elasticity moduli of the cement sheath. It can be seen that the influence on the casing's inner diameter in the two concave areas is different. In the first concave area, with the increase of the elasticity modulus of the cement sheath, the reduction of the casing's inner diameter decreases. But in the second concave area, the variation of the diameter has a reverse rule (Figure 21). Figures 22 and 23 show the influence of the Poisson ratio on the casing's inner diameter. It can be seen that with the increase of the Poisson ratio, the reduction of diameter increases, especially in the second concave area, which indicates that the lower Poisson ratio is beneficial for decreasing the reduction of the casing's inner diameter.

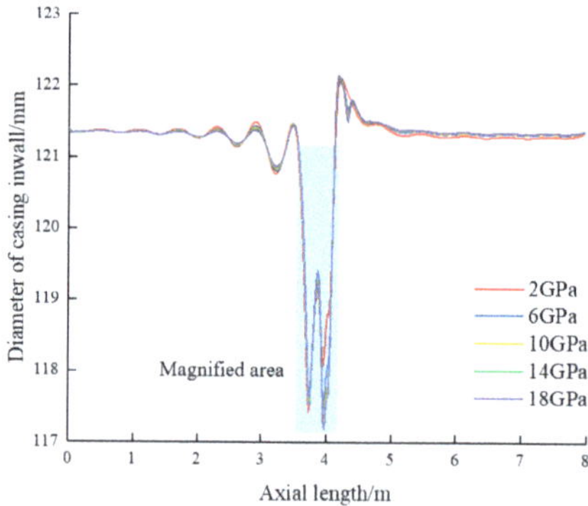

Figure 20. Variation of diameter under different elasticity moduli.

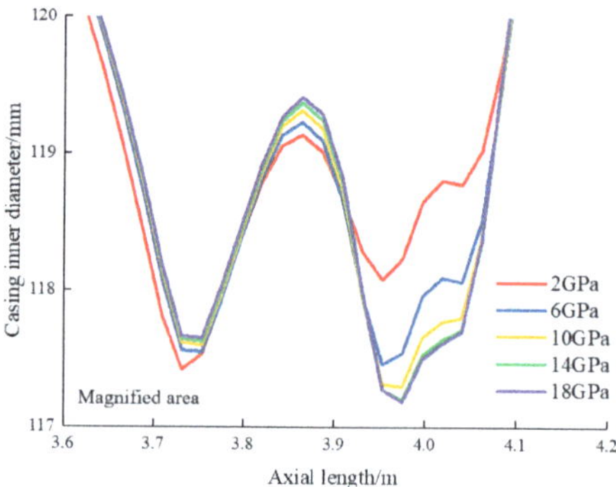

Figure 21. Magnified area.

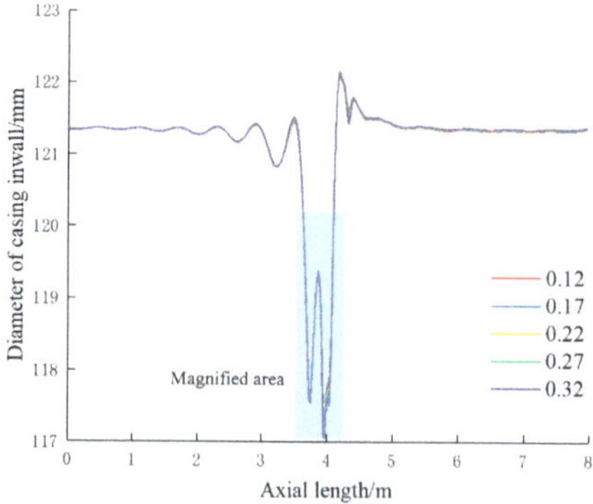

Figure 22. Variation of diameter under different Poisson ratios.

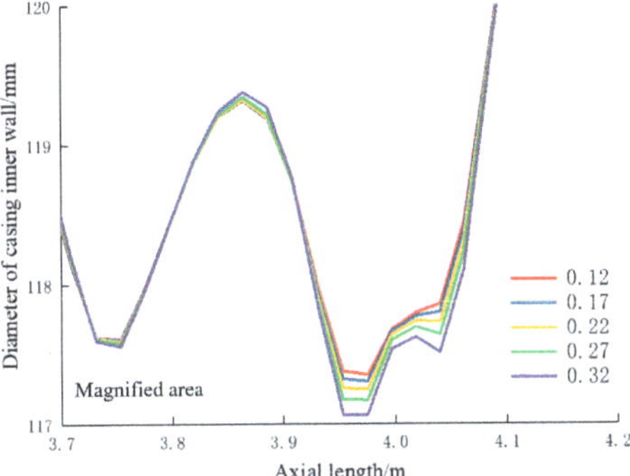

Figure 23. Magnified area.

6. Results Comparison and Mitigation Method

The method in this study presented a way to evaluate the reduction of casing's inner diameter. Using to the microseismic measurements, microseismic moment magnitude can be obtained. And based on the study of Chen et al. [14], slip distance can be calculated. Then, by using the model proposed in this study, the diameter of a casing's inner wall after fault slipping can be computed. The computed results and the measurement results by using MFC tools were compared, as shown in Figure 24. It can be seen that the numerical method in this study has an accuracy of up to 90.17%, and it can be used to select the soluble bridge plug after casing shear deformation, achieving the purpose of overcoming the problem that MFC measurement is costly, as mentioned above.

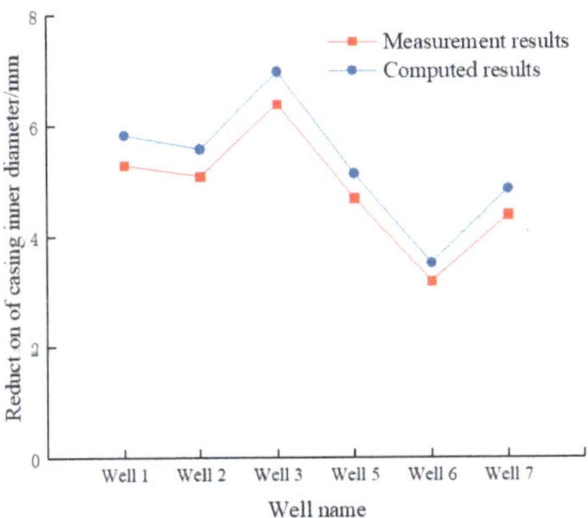

Figure 24. Comparison of measurement and computed results.

According to the above analysis, it could be seen that decreasing the slip distance was the best way to protect the casing. In order to mitigate or eliminate fault slipping, the interface between the Nisku and Ireton formations should avoid being opened during the operation of well cementation. As a consequence, the well structure was optimized. The depth of the intermediate casing shoe was reduced to approximately 3600 m, and the depth was more than 100 m above the interface between the Nisku and Ireton formations. The method was applied to nine wells, eight wells did not occur casing shear deformation, which was proved to be the most efficient and economical way. And the engineering practice supported the correctness of the analysis in this study.

7. Conclusions

Casing shear deformation occurring during multistage fracturing was monitored by using MFC tools, and the cause of the shear deformed points located at the interface between the Nisku and Ireton formations was analyzed. A new investigation based on the MFC measurement results was carried out, and the impact of influential factors on the reduction of a casing's inner diameter was studied. The following conclusions were drawn:

(1) MFC surveys were carried out to monitor the casing deformation occurring during multistage fracturing in Simonette, Canada. Statistical data showed that shear deformation was the main type of casing deformation, and the shear deformed points can be classified as two types according to the positions of occurrence: shear deformed points located at the position of the interface between the Nisku and Ireton formations (75%), and in the horizontal segment (25%).

(2) The cause of casing shear deformation occurring at the interface between the Nisku and Ireton formations was analyzed. When the interface between the different layers was opened during the operation of well cementation, the friction coefficient between the layers decreases dramatically. During multistage fracturing, the fault was activated and slipped along the interface, which was verified by the microseismic data.

(3) A numerical model has been developed to analyze the reduction of the casing's inner diameter. The simulation results showed that: (a) fault slipping caused the reduction of casing inner diameter, and the maximum change appeared at the position of the interface of the two formations; (b) the cross-section of the casing calculated by the numerical model was similar to the shape reflected by MFC data in that particular position.

(4) A sensitivity analysis was carried out and the influence of slip distance, the casing's inner pressure, the mechanical parameters of the cement sheath, and the intermediate and production casing thickness on the reduction of the casing's inner diameter were analyzed. According to the numerical analysis results, decreasing the slip distance, maintaining high pressure, decreasing the Poisson ratio of the cement sheath, and increasing the casing thickness were beneficial for protecting the integrity of casing. Furthermore, the effectiveness of increasing the intermediate casing thickness is greater than increasing that of the production casing.

(5) Measurement results were compared with computed results to verify the method proposed in this study. The numerical method in this study has an accuracy of up to 90.17%, which can be used as basis for choosing soluble bridge plugs. In addition, the well structure was optimized, and the depth of the intermediate casing shoe was reduced to approximately 3600 m, which was more than 100 m above the interface between the Nisku and Ireton formations. The effectiveness of this method was verified by engineering in practice, as eight of nine wells did not incur casing deformation after implementation of the method.

Author Contributions: Y.X. contributed to developing the mathematical model, performed the data analysis, and wrote the manuscript; J.L. (Jun Li) and G.L. supervised the research and edited the manuscript. J.L. (Jianping Li) and J.J. analyzed the engineering data.

Funding: "Study on failure mechanisms and control methods of wellbore integrity of shale gas horizontal wells" (U1762211);"Optimum research of non-uniform cluster perforation along the long horizontal section in heterogeneous shale reservoirs" (51674272).

Acknowledgments: This research was financially supported by the Key Program of National Natural Science Foundation of China "Study on failure mechanisms and control methods of wellbore integrity of shale gas horizontal wells" (U1762211), the National Natural Science Funds "Optimum research of non-uniform cluster perforation along the long horizontal section in heterogeneous shale reservoirs" (51674272).

Conflicts of Interest: The authors declare no conflict of interest.

Appendix A

The stress field of the formation was usually described by triaxial principal in-situ stresses, including the horizontal maximum in-situ stress, the horizontal minimum in-situ stress and the vertical in-situ stress (σ_H, σ_h and σ_v, MPa). Although the stress field in the vertical section of the shale gas well was stay the same with in-situ stress, the mechanical state in the inclination section was different. For the reason that the research object was located in the inclination section, the in-situ stress should be converted to the stress tensor in the wellbore coordinate system.

A coordinate rotation matrix was built in order to transform the original coordinate system to wellbore coordinate system. It could be expressed as

$$L = L_y L_z = \begin{bmatrix} \cos\alpha & 0 & -\sin\alpha \\ 0 & 1 & 0 \\ \sin\alpha & 0 & \sin\alpha \end{bmatrix} \begin{bmatrix} \cos\beta & \sin\beta & 0 \\ -\sin\beta & \cos\beta & 0 \\ 0 & 0 & 1 \end{bmatrix} = \begin{bmatrix} \cos\alpha\cos\beta & \cos\alpha\sin\beta & -\sin\alpha \\ -\sin\beta & \cos\beta & 0 \\ \sin\alpha\cos\beta & \sin\alpha\sin\beta & \cos\alpha \end{bmatrix} \quad \text{(A1)}$$

where, α represents the hole inclination angle,°; β represents the well azimuth angle,°; L_y and L_z represent the direction cosine matrixes rotating around y-axis and z-axis based on the right-hand rule.

$$L_y = \begin{bmatrix} \cos\alpha & 0 & -\sin\alpha \\ 0 & 1 & 0 \\ \sin\alpha & 0 & \sin\alpha \end{bmatrix} \quad \text{(A2)}$$

$$L_z = \begin{bmatrix} \cos\beta & \sin\beta & 0 \\ -\sin\beta & \cos\beta & 0 \\ 0 & 0 & 1 \end{bmatrix} \quad \text{(A3)}$$

Then, the stress tensor (σ_{ij}) in the wellbore coordinate could be calculated by

$$\sigma_{ij} = \begin{bmatrix} \sigma_{xx} & \sigma_{xy} & \sigma_{xz} \\ \sigma_{yx} & \sigma_{yy} & \sigma_{yz} \\ \sigma_{zx} & \sigma_{zy} & \sigma_{zz} \end{bmatrix} = L \begin{bmatrix} \sigma_H & & \\ & \sigma_h & \\ & & \sigma_v \end{bmatrix} L^T \quad \text{(A4)}$$

where, L^T represents the transposition of L, dimensionless.

Appendix B

A linear relationship was assumed between the formation temperature and depth:

$$T_z = T_b + \alpha(z - b) \quad \text{(A5)}$$

where T_z represents formation temperature at a certain depth, °C, T_b represents land surface temperature, °C, α represents the geothermal gradient, °C/m, z represents reservoir depth, m; b represents the benchmark depth, m.

Based on the law of energy conservation, a model of the wellbore temperature field during fracturing was established. The wellbore was divided into infinitesimal equal parts along the wellbore. The conservation of energy of fluid in wellbore is calculated according to Equations (A6) and (A7):

$$Q\rho_0 C_0 T_{0,j-1}^{n+1} - Q\rho_0 C_0 T_{0,j}^{n+1} + 2\pi r_0 \Delta H_j U\left(T_{1,j}^{n+1} - T_{0,j-\frac{1}{2}}^{n+1}\right) + W_j = \pi {r_0}^2 \Delta H_j \rho_0 C_0 \frac{T_{0,j-\frac{1}{2}}^{n+1} - T_{0,j-\frac{1}{2}}^{n}}{\Delta t} \quad \text{(A6)}$$

$$W_j = \lambda_{fj} \frac{\Delta H_j}{r_0} \frac{\rho_0}{2} \frac{Q^3}{\pi^2 {r_0}^2} \quad \text{(A7)}$$

$$\frac{T_{0,j}^{n+1} - T_{0,j-1}^{n+1}}{2} = T_{0,j-\frac{1}{2}}^{n+1}$$

The conservation of energy of solid elements with a fluid foundation and for solid elements was shown in Equations (A8) and (A9), respectively:

$$-2\pi r_0 \Delta H_j U\left(T_{1,j}^{n+1}-T_{0,j-\frac{1}{2}}^{n+1}\right)+2\pi r_1 \Delta H_j K_1 \frac{T_{2,j}^{n+1}-T_{1,j}^{n+1}}{\frac{r_2-r_0}{2}}=\pi\left(r_1{}^2-r_0{}^2\right)\Delta H_j \rho_1 C_1 \frac{T_{1,j}^{n+1}-T_{1,j}^{n}}{\Delta t} \quad \text{(A8)}$$

$$\frac{4\pi r_{i-1}\Delta H_j K_{i-1}}{r_i-r_{i-2}}T_{i-1,j}^{n+1}+\left(-\frac{4\pi r_{i-1}\Delta H_j K_{i-1}}{r_i-r_{i-2}}-\frac{4\pi r_i \Delta H_j K_i}{r_{i+1}-r_{i-1}}-\frac{\pi\left(r_i{}^2-r_{i-1}{}^2\right)\Delta H_j \rho_i C_i}{\Delta t}\right)T_{i,j}^{n+1}+\frac{4\pi r_i \Delta H_j K_i}{r_{i+1}-r_{i-1}}T_{i+1,j}^{n+1}=-\frac{\pi\left(r_i{}^2-r_{i-1}{}^2\right)\Delta H_j \rho_i C_i}{\Delta t}T_{i,j}^{n+1} \quad \text{(A9)}$$

where W_j represents the heat generated by friction between the fracturing fluid and casing wall, J; Q represents the displacement of fracturing fluid, m^3/min, ρ represents density, kg/m^3, C represents specific heat, $J/(kg \cdot {}^\circ C)$, r is the radius, m; ΔH_j represents the height of the control unit body, m; U represents the convective heat transfer coefficient between the fracturing fluid and casing wall, $w/(m^2 \cdot {}^\circ C)$; and λ_{fj} represents the casing friction coefficient, dimensionless.

During the progress of meshing, when $i = 0$, $1 \le i < m$, $m \le i < n$, $n \le i < o$, $o \le i < p$, $p \le i < q$, the meshing grids represent the fracturing fluid, production casing, cement sheath (1), intermediate casing, cement sheat (2), and formation, as show in Figure A1.

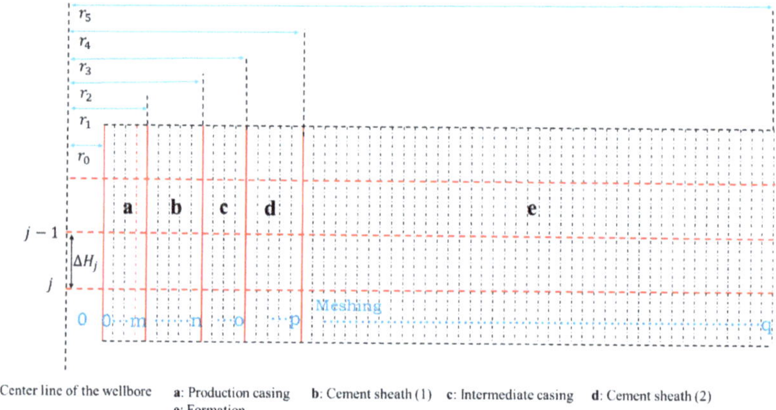

Figure A1. Meshing grids of the assembly.

When the fracturing fluid was pumped into the wellbore downhole with high discharge, it was in the state of turbulence, The convective heat transfer coefficient between the casing and fluid can be calculated by using the Marshall model shown in Equation (A10) [34].

$$U=\frac{S_t K_0}{D}=0.0107\frac{K_0}{D}\left\{\rho_0 D_{eff}\frac{4Q}{\pi D^2}/\left[K\left(\frac{3n+1}{4n}\right)^n\left(\frac{32Q}{\pi D^3}\right)^{n-1}\right]\right\}^{0.67}\left[K_{con}\left(\frac{3n+1}{4n}\right)^n\left(\frac{32Q}{\pi D^3}\right)^{n-1}C_0/K_0\right]^{0.33} \quad \text{(A10)}$$

where S_t is the Stanton number (dimensionless), K_0 is the heat conductivity coefficient ($w/(m \cdot {}^\circ C)$; D is the casing diameter (m); D_{eff} is the equivalent diameter of casing (m); n is the liquidity index, dimensionless; K_{con} is the consistency (Pa/s^n); C_0 represents the specific heat of the fracturing fluid ($J/(kg \cdot {}^\circ C)$).

According to Equations (A5)–(A10), changes in the temperature of the fracturing fluid can be obtained by calculation, and the temperature history of any position along the horizontal section can be expressed as functions that are used as the basic parameters in the finite element model:

$$T_l^Q = g(Q, l, t) \quad \text{(A11)}$$

where l is the distance between the selected position and the toe-end (m).

References

1. Meyer, B.R.; Bazan, L.W. A Discrete Fracture Network Model for Hydraulically Induced Fractures-Theory, Parametric and Case Studies. In Proceedings of the SPE Hydraulic Fracturing Technology Conference, Woodlands, TX, USA, 24–26 January 2011.
2. Wu, Q.; Xu, Y.; Zhang, S.; Wang, T.; Guan, B.; Wu, G.; Wang, X. The Core Theories and Key Optimization Design of Volume Simulation Technology for Unconventional Reservoirs. *Acta Petrolei Sin.* **2014**, *35*, 706–714.
3. Zhou, D.; Zheng, P.; He, P.; Peng, J. Hydraulic Fracture Propagation Direction during Volume Fracturing in Unconventional Reservoirs. *J. Petrol. Sci. Eng.* **2016**, *141*, 82–89. [CrossRef]
4. Wang, F.; Li, B.; Zhang, Y.; Zhang, S. Coupled Thermo-Hydro-Mechanical-Chemical Modeling of Water Leak-Off Process during Hydraulic Fracturing in Shale Gas Reservoirs. *Energies* **2017**, *10*, 1960. [CrossRef]
5. Tao, H.; Zhang, L.; Liu, Q.; Qi, D.; Luo, M.; Zhao, Y. An Analytical Flow Model for Heterogeneous Multi-Fractured Systems in Shale Gas Reservoirs. *Energies* **2018**, *11*, 3422. [CrossRef]
6. Bao, X.; David, W.E. Fault Activation by Hydraulic Fracturing in Western Canada. *Science* **2016**, *17*, 1–9. [CrossRef] [PubMed]
7. Nagel, N.B.; Garcia, X.; Sanchez Nagel, M.A.; Lee, B. Understanding "SRV": A Numerical Investigation of "Wet" vs. "Dry" Microseismicity during Hydraulic Fracturing. In Proceedings of the SPE Annual Technical Conference and Exhibition, San Antonio, TX, USA, 8–10 October 2012.
8. Lian, Z.; Yu, H.; Lin, T.; Guo, J. A Study on Casing Deformation Failure during Multi-Stage Hydraulic Fracturing for the Stimulated Reservoir Volume of Horizontal Shale Wells. *J. Nat. Gas Sci. Eng.* **2015**, *23*, 538–546. [CrossRef]
9. Jalali, H.H.; Rofooei, F.R.; Attarin, N.K.A.; Samadianc, M. Experimental and Finite Element Study of the Reverse Faulting Effects on Buried Continuous Steel Gas Pipelines. *Soil Dyn. Earthq. Eng.* **2016**, *86*, 1–14. [CrossRef]
10. Adams, N.J.; Mitchell, R.F.; Eustes, A.W.; Sampaio, J.H.; Antonio, A.O. A Causation Investigation for Observed Casing Failures Occurring during Fracturing Operations. In Proceedings of the SPE Hydraulic Fracturing Technology Conference and Exhibition, Woodlands, TX, USA, 24–26 January 2017.
11. Meyer, J.J.; Gallop, J.; Chen, A.; Reynods, S.; Mildren, S. Can Seismic Inversion Be Used for Geomechanics? A Casing Deformation Example. In Proceedings of the Unconventional Resources Technology Conference, Houston, TX, USA, 23–25 July 2018.
12. Saada, M.; Ismail, M.; Lu, Y.; Zhang, W.; Moula, M. Measure The Effectiveness of Frac Sleeves Integrity in Multi-Stage Completions Using Multi-Finger Caliper and Acoustic Flow-Analyzer. In Proceedings of the SPE Kingdom of Saudi Arabia Annual Technical Symposium and Exhibition, Damman, Saudi Arbia, 23–26 April 2018.
13. Simon, S.; Husam, A.R.; Basant, K.S.; Graeme, D. OCTG Corrosion Status Monitoring in Wells Completed with Challenging High Ni CRA Non-Magnetic OCTG in Gas Well Environment with Higher Temperature, Extreme Hydrogen Sulphide and Reservoir Conditions–A Case Study. In Proceedings of the Abu Dhabi International Petroleum Exhibition & Conference, Abu Dhabi, UAE, 12–15 November 2018.
14. Yan, W.; Zou, L.; Li, H.; Deng, J.; Ge, H.; Wang, H. Investigation of Casing Deformation During Hydraulic Fracturing in High Geo-Stress Shale Gas Play. *J. Petrol. Sci. Eng.* **2017**, *150*, 22–29. [CrossRef]
15. Xi, Y.; Li, J.; Liu, G.; Cha, C.; Fu, Y. Numerical Investigation for Different Casing Deformation Reasons in Weiyuan-Changning Shale Gas Field During Multistage Hydraulic Fracturing. *J. Petrol. Sci. Eng.* **2018**, *163*, 691–702. [CrossRef]
16. Chen, Z.; Wang, P.; Xiang, D. Analysis of Casing Deformation in the Changning-Weiyuan Block Based on Focal Mechanism. *Petrol. Drill. Tech.* **2017**, *45*, 110–114.
17. Noshi, C.I.; Noynaert, S.F.; Schubert, J.J. Casing Failure Data Analytics: A Novel Data Mining Approach in Predicting Casing Failures for Improved Drilling Performance and Production Optimization. In Proceedings of the SPE Annual Technical Conference and Exhibition, Dallas, TX, USA, 24–26 September 2018.
18. Chen, Z.; Shi, L.; Xiang, D. Mechanism of Casing Deformation in the Changning–Weiyuan National Shale Gas Project Demonstration Area and Countermeasures. *Nature Gas Ind.* **2016**, *36*, 70–75.
19. Yin, F.; Deng, Y.; He, Y.; Gao, D.; Hou, B. Mechanical Behavior of Casing Crossing Slip Formation in Water Flooding Oilfields. *J. Petrol. Sci. Eng.* **2018**, *167*, 796–802. [CrossRef]

20. Qian, B.; Yin, C.; Li, Y.; Xu, B.; Qin, G. Diagnostics of Casing Deformation in Multi-Stage Hydraulic Fracturing Stimulation in Lower Silurian Marine Shale Play In Southwestern, China. In Proceedings of the Unconventional Resources Technology Conference, San Antonio, TX, USA, 20–22 July 2015.
21. Zoback, M.D.; Snee, J.E.L. Predicted and observed shear on pre-existing faults during hydraulic fracture stimulation. In Proceedings of the SEG International Exposition and 88th Annual Meeting, Anaheim, CA, USA, 14–19 October 2018.
22. Marc, M. Monitoring Shear Deformations above Compacting High-Pressure High-Temperature Reservoirs With Calliper Surveys. *Int. J. Rock Mech. Min. Sci.* **2016**, *86*, 292–302.
23. Gao, L.; Qiao, L.; Liu, Z.; Zhuang, Z.; Yang, H. Numerical Modeling and Cementing Countermeasure Analysis of Casing Shear Damage in Shale Reservoir. *China Petrol. Mach.* **2016**, *44*, 6–10.
24. Guo, X.; Li, J.; Liu, G.; Zeng, Y.; He, M.; Yan, H. Numerical Simulation of Casing Deformation During Volume Fracturing of Horizontal Shale Gas Wells. *J. Petrol. Sci. Eng.* **2019**, *172*, 731–742. [CrossRef]
25. Davis, M.; Karlen, G.A. Regional Assessment of the Duvernay Formation; A World-Class Liquids-Rich Shale Play. In Proceedings of the GeoConvention 2013: Integration, Calgary, AB, Canada, 6–10 May 2013.
26. Rodgers, B. Declining Costs Enhance Duvernay Shale Economics. *Oil Gas J.* **2014**, *112*, 70.
27. Fox, A. A Regional Geomechanical Study of the Duvernay Formation in Alberta, Canada. In Proceedings of the GeoConvention 2015: New Horizon, Calgary, AB, Canada, 4–8 May 2015.
28. Zoback, M.D. *Reservoir Geomichanics*, 2nd ed.; Cambridge University Press: London, UK, 2010; pp. 43–95.
29. Biot, M.A. General Theory of three-dimensional consolidation. *J. Appl. Phys.* **1941**, *12*, 155–164. [CrossRef]
30. Goodman, R.E. *Introduction to rock mechanics.*, 2nd ed.; Wiley: Chichester, UK, 1989; pp. 21–78.
31. Brune, J. Tectonic Stress and the Spectra of Shear Waves from Earthquakes. *J. Geophys. Res.* **1970**, *75*, 4997–5009. [CrossRef]
32. Hanks, T.C.; Kanamori, H. A Moment Magnitude Scale. *J. Geophys. Res.* **1979**, *84*, 2348–2350. [CrossRef]
33. Cipolla, C.L.; Mack, M.G.; Maxwell, S.C.; Downie, R.C. A Practical Guide to Interpreting Microseismic Measurements. In Proceedings of the North American Unconventional Gas Conference and Exhibition, Woodlands, TX, USA, 14–16 June 2011.
34. Marshall, D.W.; Bentsen, R.G. A computer model to determine the temperature distributions in a wellbore. *J. Can. Petrol. Technol.* **1982**, *21*, 63–75. [CrossRef]

© 2019 by the authors. Licensee MDPI, Basel, Switzerland. This article is an open access article distributed under the terms and conditions of the Creative Commons Attribution (CC BY) license (http://creativecommons.org/licenses/by/4.0/).

Article

A Data-Driven Workflow Approach to Optimization of Fracture Spacing in Multi-Fractured Shale Oil Wells

Xu Yang [1,*] and Boyun Guo [2,*]

1 School of Oil & Gas Engineering, Southwest Petroleum University, Chengdu 610500, China
2 Department of Petroleum Engineering, University of Louisiana at Lafayette, Lafayette, LA 70504, USA
* Correspondence: yangxuzq@sina.com (X.Y.); guo.boyun@gmail.com (B.G.)

Received: 2 April 2019; Accepted: 21 May 2019; Published: 23 May 2019

Abstract: A data-driven workflow approach is presented in this study for optimizing fracture spacing of multifractured horizontal wells (MFHW) in shale oil reservoirs. The workflow employs a simple well productivity model for the initial design of hydraulic fracturing well completions. This provides a transparent approach to the identification of key fracturing parameters affecting well productivity. The workflow uses transient pressure or production data to identify fracture interference. This offers a reliable and cost-effective means for assessment of well production potential in terms of optimization of fracture spacing in the MFHW. Result of a field case study indicated that three wells were drilled in an area with dense natural fractures, and the fracture spacing of MFHW in this area was short enough to effectively drain the stimulated reservoir volume (SRV), while the other three wells were drilled in an area with less natural fractures, and the fracture spacing of MFHW in this area could be shortened to double well productivity.

Keywords: unconventional; shale oil; well; fracturing; optimization

---

1. Introduction

Thanks to multifractured horizontal well (MFHW) technology, development of shale oil fields in the past five years has made United States energy self-sufficient. However, the results of the technology in other regions of the world are mixed. This is partially attributed to inadequate effort in optimizing a completed design of MHFW.

Modern MFHW techniques include simul frac, zipper frac, and modified zipper frac. Simul frac creates simultaneous hydraulic fractures in symmetric stages in two wells. Zipper frac creates alternate hydraulic fractures in symmetric stages in two wells. Modified zipper frac creates hydraulic fractures in nonsymmetric stages in two alternate wells [1]. These fracturing techniques create symmetric, transverse fractures of ideally the same length and spacing. This common structure of multifractured horizontal wells allows for using general mathematical models to predict well productivity in the stage of well completion design.

The available methods for predicting shale well productivity include analytical transient flow models [2–6] and numerical computer models [7–11]. Analytical transient flow models are mostly used in pressure transient test analyses rather than in optimizing well completion design. Numerical computer models are flexible in handling systems with nonsymmetrical fractures and multiphase flows [12]. In addition to model complicity, the uncertainty in locations of natural fractures is a concern regarding the accuracy of computer simulation results. Based on the work of Guo et al. [13], Zhang et al. [14] presented an analytical solution for estimating long-term productivity of multifractured shale gas wells by coupling the linear flow in the reservoir and the linear flow in hydraulic fractures. Li et al. [15] extended the analytical solution to shale oil wells and validated the solutions with field data from an oil well and a gas well.

The results of MFHW technology in different regions of the world are inconsistent. This is believed partially because of the lack of optimization of MHFW parameters, especially fracture spacing. Optimization of such spacing is considered as one of the most critical steps in enabling economic of horizontal wells and requires a lot of attention [7]. A data-driven workflow procedure, combined with simple well productivity models, is presented in this paper for fracture spacing optimization. This workflow procedure has the advantage of using a simple analytical well productivity model driven by real production data, making it a practical approach to optimize multistage fracturing of horizontal wells in shale oil reservoirs.

2. Concept of Fracture Spacing

Modern shale oil and gas horizontal wells are mostly fractured in multiple stages utilizing plug and perf methodology, where the stages are separated by frac plugs set inside the casing. Perforation clusters are created for each stage. The explosive energy during perforating should induce microcracks from the perforations [16]. When fracturing fluid is forced to go through the perforations, multiple fractures may be initiated from these microcracks or other heterogeneities such as natural fractures, drilling-induced fractures, rock texture, orientation and magnitude of formation stresses, etc. The orientations of fractures near the wellbore area are complicated because the wellbore and perforations alter the state of stress in this area [16].

A number of fracture propagation mechanisms may exist. Fractures may re-orientate from the perforations towards preferred fracture planes, owing to stress anisotropy. Fractures may also interact with and merge to natural fractures if the latter exists. As a result, multiple and tortuous fractures are anticipated near the wellbore area. When these initial fractures propagate away from the wellbore area, they may converge or diverge depending on stress field and natural fractures. Branching due to shale heterogeneities may develop from a single fracture. All fractures initiated from a perforation cluster should form a hydraulic fracture in the plane normal to the minimum principal stress to achieve the minimum strain energy in the shale. Twelve hydraulic fractures created from 12 perforation clusters are depicted in Figure 1. The fracture spacing is defined in this study as the distance between the centerlines of two adjoined hydraulic fractures, which are expected to be equal to the distance between the two adjoined perforation clusters.

Although well productivity models, such as those presented by Zhang et al. [14] and Li et al. [15], suggest reducing fracture spacing for maximizing well productivity, short spacing may cause early interferences between fractures as a result of fracture branching. Whether early interference exists or not depends on local geological conditions, especially natural fractures and local rock stress fields. Fracture interference can be identified by pressure transient data analyses and/or production rate transient data analyses.

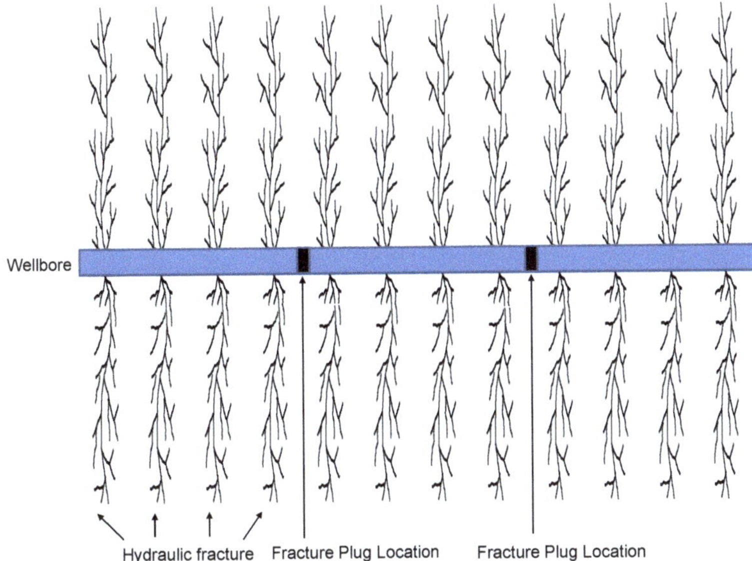

Figure 1. Twelve hydraulic fractures developed from 12 perforation clusters in three stages of fracturing.

3. Workflow for Optimization

The proposed workflow to optimize fracture spacing is based on industry practice [16], with the addition of data analysis, to identify fracture interference. It is summarized in the following steps:

1. Design a fracturing job for the first well in the area for a desirable production rate using a well productivity model. Fracture spacing is selected based on horizontal wellbore length, volumes of fracturing fluid and proppant, and completion tools.
2. Execute the fracturing job using the designed fracture spacing and other parameters.
3. Run a pressure transient test on the well if possible, and put the well into production.
4. Perform transient pressure or transient production rate data analyses to identify fracture interference.
5. If well completion permits, refracture the well based on the identified fracture interference.
6. Modify the well completion design including the fracturing treatment design and/or the spacing between perforation clusters for the next well on the basis of fracture interference in the previous well.

Some of these workflow steps are further outlined in the subsections that follow.

3.1. Fracturing Design

The physics of fluid flow in shale oil reservoirs is adequately described in the literature. The assumptions employed in this study were oversimplified, as the desorption of hydrocarbon from rock matrix and non-Darcy flow were not considered. However, it is a fact that pressure transient and production data support the mathematical models derived from Darcy's law. This implies that fluid flow in shale oil reservoirs is dominated by Darcy's law.

Fracturing design can be guided by mathematical models derived from Darcy's law for shale gas and oil well production under pseudosteady flow conditions. For multifractured horizontal oil wells, Li et al.'s model for oil production rate is expressed as [15]:

$$q_o = \frac{5.91 \times 10^{-3} n_f k_m h_f (\overline{p} - p_w)}{B_o \mu_o S_f \sqrt{c}\left(\frac{1}{1-e^{-\sqrt{c}x_f}} - \frac{1}{3x_f\sqrt{c}}\right)}, \quad (1)$$

where c is expressed as:

$$c = \frac{96k_m}{k_f w S_f}, \tag{2}$$

where q_o is the well production rate in stb/d, n_f is the number of perforation clusters (hydraulic fractures), k_m is the matrix permeability in md, h_f is the fracture height in ft, $\overline{p}$ is the average formation pressure in psia, p_w is the wellbore pressure in psia, μ_o is the oil viscosity in cp, B_o is the oil formation factor in rb/stb, S_f is the fracture spacing in ft, e is an exponential function, x_f is the hydraulic fracture half-length in ft, k_f is the fracture permeability in md, and w is the average fracture width in inches.

Substituting Equation (2) into Equation (1) yields:

$$q_o = \frac{6.03 \times 10^{-4} n_f h_f (\overline{p} - p_w)}{B_o \mu_o \left(\frac{1}{1-e^{-\sqrt{c}x}} - \frac{1}{3x_f\sqrt{c}} \right)} \sqrt{\frac{k_m k_f w}{S_f}}. \tag{3}$$

Although Equation (3) shows that maximum well productivity can be obtained by optimizing the effects of the number of hydraulic fractures, hydraulic fracture spacing, and hydraulic fracture width, this is not true in reality as fracture complexity and additional factors such as frac hits also exist. As pointed out by Potapenko et al. [17], in most of the unconventional reservoirs where the majority of multistage wells are drilled, the fracture geometry is quite complex and interaction between fractures starts very early. In this case, spacing between perforation clusters should not have much impact on well productivity. However, in situations where a significant portion of hydraulic fractures is damaged during well startup, resulting in disconnections of such fractures from the wellbore, having small spacing between perforation clusters still appears to be beneficial because disconnected fractures still may contribute to production through other nondamaged perforation clusters [17]. The issues of fracture complexity, interaction, and disconnection are quite difficult to consider in mathematical modeling because there is a lack of data to explicitly describe their configurations and dynamics [18]. These complication issues are assumed to be negligible in the analysis that follows.

For a given horizontal wellbore length L, well operators tend to increase the number of perforation clusters during the well completion optimization process. It brings some additional considerations about fracture length, the number of frac stages, and the volume of frac treatment. The number of clusters is expressed as:

$$n_f = \frac{L}{S_f}. \tag{4}$$

For a given total volume of fracture proppant V_f in ft^3, the following material balance holds:

$$V_f = n_f (2x_f) \left(\frac{w}{12} \right) h_f. \tag{5}$$

It should be mentioned that w in Equation (4) is in inches, and V_f is the bulk volume (not the physical volume) of proppant.

Substituting Equation (4) into Equation (5) and rearranging the latter gives:

$$w = \frac{6 V_f S_f}{x_f h_f L}. \tag{6}$$

Substituting Equations (4) and (6) into Equation (3) results in:

$$q_o = \frac{1.48 \times 10^{-3} (\overline{p} - p_w)}{B_o \mu_o S_f \left(\frac{1}{1-e^{-\sqrt{c}x_f}} - \frac{1}{3x_f\sqrt{c}} \right)} \sqrt{\frac{k_m k_f h_f V_f L}{x_f}}. \tag{7}$$

The average reservoir pressure is [15]:

$$\overline{p} = p_e - \frac{p_e - p_w}{3x_f\sqrt{c}}\left(1 - e^{-\sqrt{c}x_f}\right), \tag{8}$$

where p_e is the reservoir pressure in psia.

It can be shown that in the practical ranges of parameters the sum of the two terms in the bracket in the denominator is very close to unity. Therefore, if fracture spacing is reduced, both the number of clusters and the total fracture surface will increase. Under the assumption of reservoir linear flow (RLF) with infinite fracture conductivity, the well production rate is proportional to the total fracture surface area. This supports the concept of massive volume fracturing where many clusters with the shortest possible spacings are used for pumping massive proppant into the created hydraulic fractures.

Although cluster spacing should be theoretically as low as possible for maximizing well productivity, a minimum required cluster spacing (MRCS) has to be considered in practice if temporal change and well lifetime are not considered. MRCS is controlled by (1) well completion design and equipment limitations and (2) hydraulic fracture interference. In well completion design, frac plugs and perforation clusters themselves are spaced apart, and casing couplings have to be avoided in perforation clusters. Perf gun can carry a limited number of charges, which are used for the creation of perf clusters. Adding a higher number of perf clusters may require a higher number of perf trips, which impacts overall cost and the operational efficiency. For the concern of hydraulic fracture interferences, sever connections of hydraulic fractures from different perforation clusters are usually not desirable, although overlapping of the frac networks from different perforation clusters may be quite beneficial to well productivity. This interference can be assessed using transient pressure or transient production data analysis.

3.2. Transient Pressure Analysis for Fracture Interference Identification

The theoretical basis of pressure transient data analysis can be found in numerous references [19–21]. Modern computer software packages available for pressure transient and production data analyses include F.A.S.T. WellTest [22] and PanSystem [23]. Applications of the pressure transient data analysis theory to shale gas and oil wells are shown by a number of investigators such as Shan et al. [24], Pang et al. [25], and He et al. [26]. Figure 2 presents a diagnostic plot of pressure transient data for flow regime identification where the vertical axis is the pressure derivative defined by:

$$p' = \frac{d\Delta p}{d\ln t}, \tag{9}$$

where Δp is the pressure change, defined as reservoir pressure minus flowing bottom hole pressure for drawdown tests and shut-in bottom pressure minus the last flowing bottom hole pressure before shutting-in for pressure buildup tests. t is the test time defined as the flow time for drawdown tests and shut-in time for pressure buildup tests.

Mathematical models describing reservoir transient linear flow to fractures are found in the literature [27,28]. For a well with infinite conductivity fractures, reservoir linear flow (RLF) can be identified by the half-slope of the pressure derivative data versus time data plotted on a log–log scale. For a well with finite conductivity fractures, a bilinear flow may occur in the fracture and in the formation matrix during the initial stages. Pressure derivative data versus time data plotted on a log–log scale should show a straight line with a slope of 0.25 [29]. After the pressure change propagates to the midline between fractures, pressure derivative data versus time data plotted on a log–log scale should show a straight line with the unit slope for boundary dominated flow (BDF) during the "depletion" in the fractured volume. If a slope value greater than 0.5 is observed soon after wellbore storage, interference between fractures is indicated.

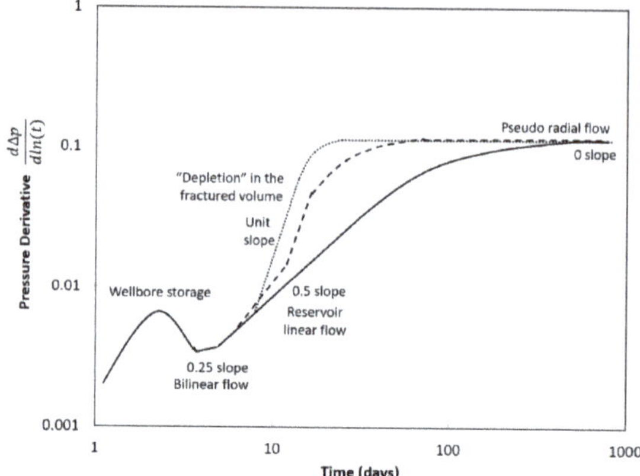

Figure 2. Diagnostic plot of pressure transient data for flow regime identification.

3.3. *Transient Rate Analysis for Fracture Interference Identification*

Applications of the transient production data analysis theory to shale oil wells are also demonstrated by previous investigators [30–32]. Figure 3 shows a diagnostic plot of rate transient data for flow regime identification. For a well with infinite conductivity fractures, the production rate data versus time data plotted on a log–log scale should show a straight line with a slope of −0.5 during the RLF. For a well with finite conductivity fractures, a bilinear flow may occur in the fracture and in the formation matrix during the initial stages. Rate data versus time data plotted on a log–log scale should show a straight line with a slope of −0.25. After the depth of investigation propagates to the midline between fractures, rate data versus time data plotted on a log–log scale should show a straight line with a slope of −1 for BDF. If a slope value between −0.5 and −1 is observed early on, interference between fractures is indicated.

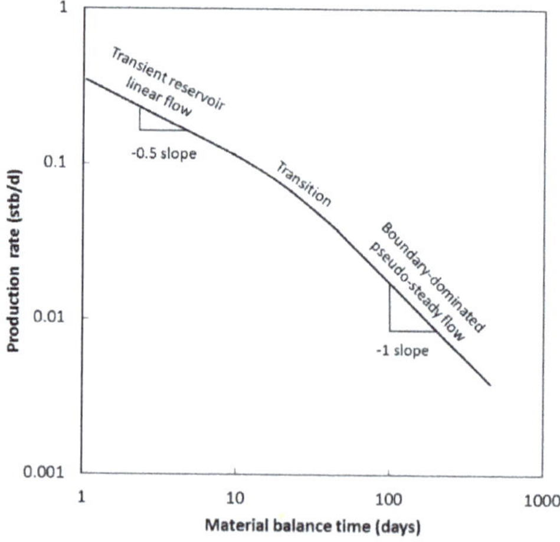

Figure 3. Diagnostic plot of rate transient data for flow regime identification.

4. Field Case Studies

The Tuscaloosa Marine Shale (TMS) formation, deposited during the Upper Cretaceous, extends from Louisiana to the southern portion of the Mississippi. Little Creek Field is located in Lincoln and Pike Counties, southwestern Mississippi, on the south rim of the Mississippi Salt Basin. It is within the Upper Cretaceous Mid-Dip Tuscaloosa trend, which occurs updip of the Lower Cretaceous shelf margin. In the Mississippi, production from the Mid-Dip trend extends 150 mi (240 km) to the east-southeast from the Mississippi River in a belt 30 to 60 mi (50–100 km) wide. Reservoir rocks are fine- to medium-grained sublitharenite in the stratigraphic unit of the Lower Tuscaloosa formation in the Late Cretaceous, Cenomanian, deposited in the fluvial meander belt environment. The productive facies are point bars. Rock total porosity is 21% to 27% with 13% to 15% primary, 3% to 4% secondary dissolution, and 5% to 8% microporosity. Core porosity ranges 10% to 35% with an average value of 24%. The core permeability measured with air ranges from 0.1 to 100 md with an average value of 10 md. The initial water saturation is between 40% and 75%, averaging at 55%. The reservoir is composed of Q and Q2 sandstone members at an average depth of 10,770 ft (3283 m) in an area of 6×3 mi^2 (9.7×4.8 km^2). The productive area is 6200 acres (2510 ha.) The hydrocarbon column height is 100 ft (30.5 m) with oil-water at 10,390 ft (3167 m) subsea. The gross sandstone thickness ranges from 15 to 85 ft (4.6–25.9 m), averaging at 40 ft (12.2 m). The average net sandstone thickness is 30 ft (9.1 m). The original reservoir pressure is 4840 psi (3.3×10^4 kPa) at 10,340 feet (3152 m) subsea. The oil has an API gravity of 39° with a GOR (gas/oil ratio) of 555:1 scf/stb and formation volume factor of 1.32. Oil viscosity is up to 5 cp (5.0×10^{-3} Pa·s) at 200 °F (93 °C).

TMS needs to use multifractured horizontal wells to become rejuvenated into a key exploration target for the industry as a leading oil field. Production rate and cumulative production data in May 2018 were gathered from 80 TMS wells in Louisiana and Mississippi. The majority of TMS wells were characterized by transient production behaviors.

Selection of wells for analysis in this study was based on the bubble map of the initial well productivity shown in Figure 4. Wells were selected from the Louisiana side of TMS with analyzable production decline curves. Only wells with initial oil production rates between 300 and 1200 stb/d were selected so wells with severe formation damage could be excluded. Wells with abnormal water cuts and gas-oil ratios were also excluded.

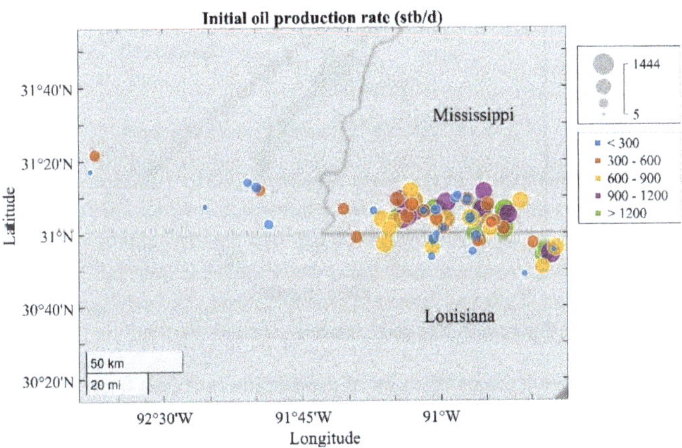

Figure 4. Bubble map of the initial well productivity of Tuscaloosa Marine Shale (TMS) wells.

Figure 5 shows the statistical results of slopes of 44 TMS wells that were used to identify RLF. It should be mentioned that 11 TMS wells were not analyzed owing to their abrupt changes in production rate over time. We saw that the slopes of 15 wells were in the range of −0.52 and −0.48. Based on the log–log plot of data from 44 oil wells in the TMS trend, a slope of −0.5 was observed for some wells but not all wells. The correlation coefficient between the initial production rate and the slope was −0.076, which meant that there was no linear correlation.

Figure 9 illustrates the calculated well productivity curve generated using the data as input to Equation (7). It showed that the model-predicted well production rate at the beginning of BDF was 121 stb/d (at S_f =69 ft), which was 0.83% higher than the observed value of 120 stb/d. The production rate at the beginning of BDF of Well #1 for S_f = 15 ft was 89% higher than that for S_f = 69 ft. This was explained by Wattenbarger's solution for the initial production rate [34]. It demonstrated that well production rate in the RLF period was proportional to the total fracture surface area. For a given horizontal wellbore length, the total fracture surface is dependent on the number of hydraulic fractures. Therefore, decreasing fracture spacing can increase production rate if the fracture half-length is held constant. Similarly, this indicated that the model-calculated well production rate at the beginning of BDF of Well #2 was 94 stb/d (at S_f = 63 ft), which was 2.08% lower than the observed value of 96 stb/d. When the fracture spacing was reduced to 15 ft, the production rate at the beginning of BDF should be 71% higher than that the original one, assuming no other complex factors were missing in the model development. Of course, the decrease in perf cluster spacing may have less than ideal impacts on well productivity as a result of other complex factors that were not considered in the model development.

Table 1. Reservoir and well completion data.

Parameters	Well #1	Well #2	Well #3
Matrix permeability	0.000105 md	0.000105 md	0.000112 md
Porosity	0.08	0.08	0.08
Slope of square root time curve	0.00032	0.00042	0.00035
Pay zone thickness	135 ft	135 ft	135 ft
Reservoir pressure	6127 psia	5876 psia	6367 psia
Wellbore pressure	4124 psi	3955 psi	4286 psi
Oil formation volume factor	1.3 rb/stb	1.3 rb/stb	1.3 rb/stb
Oil viscosity	0.5 cp	0.5 cp	0.5 cp
Total compressibility	0.00001 psi^{-1}	0.00001 psi^{-1}	0.00001 psi^{-1}
Number of perforation clusters	116	76	96
Perforation cluster spacing	69 ft	63 ft	58 ft
Hydraulic fracture half-length	315 ft	379 ft	321 ft
Volume of proppant	19,090,000 lbs	15,050,000 lbs	161,110,000 lbs
Time at the end of linear flow	180 d	150 d	120 d

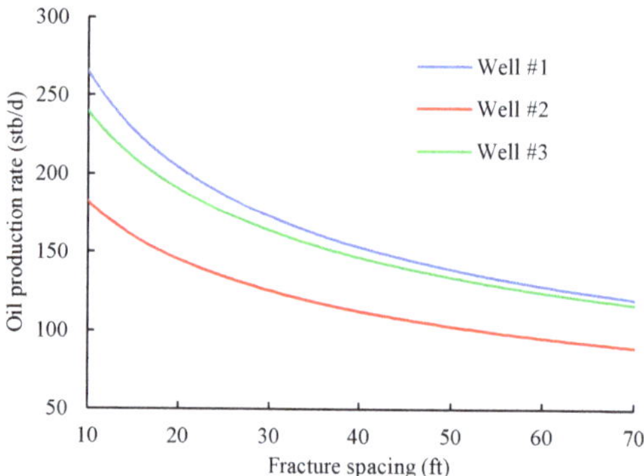

Figure 9. Model-calculated productivity for TMS wells.

Figure 10 demonstrates the diagnostic plot of production rate data for Well #4. Early production data showed a trend with a slope of −0.57, which was between −0.5 and −1, not indicating RLF. This could be due to the existence of natural fractures or the interference of hydraulic fractures. Figures 11 and 12 present the diagnostic plot of production rate data for Well #5 and Well #6, respectively. Early production data showed trends with slope values of −0.58 and −0.62, respectively, not indicating RLF. This could be due to the existence of natural fractures or the interference of hydraulic fractures. The bubble map in Figure 4 was compared with geological maps for natural fracture/shale quality identification. A further examination of locations of Wells #4, #5, and #6 confirmed that these wells were drilled in an area where multiple natural fractures were found. It was generally believed that not all the hydraulic fractures were equal in length, owing to heterogeneity within the reservoir or stress shadow. Ambrose et al. [33] indicated fracture interference had a great effect on the performance of a well. They demonstrated that fracture interference tended to increase the cumulative production. However, the higher the heterogeneity, the less the recovery factor. It was therefore not recommended to reduce fracture spacing for these wells or wells in the areas.

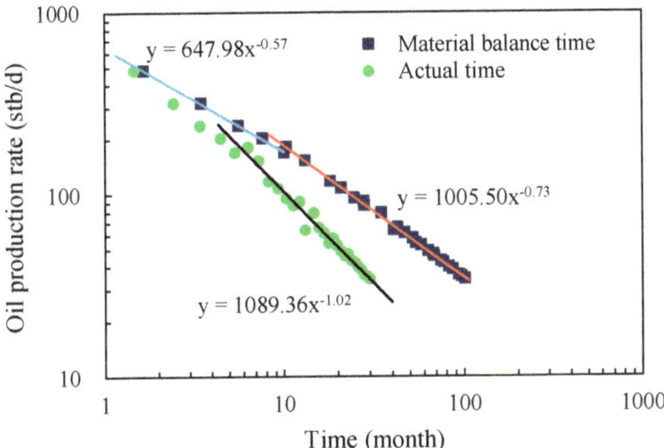

Figure 10. Plot of production rate data for Well #4.

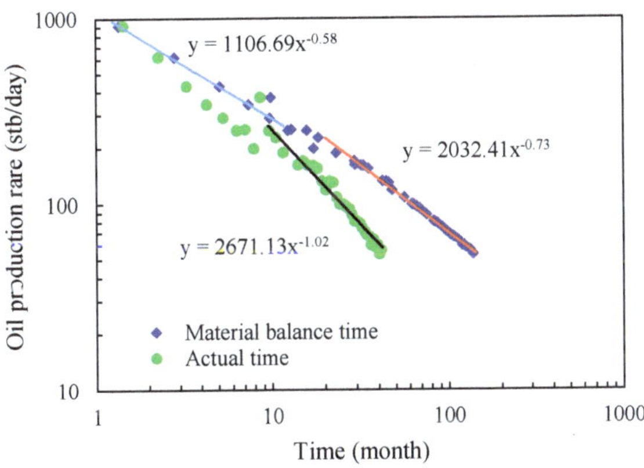

Figure 11. Plot of production rate data for Well #5.

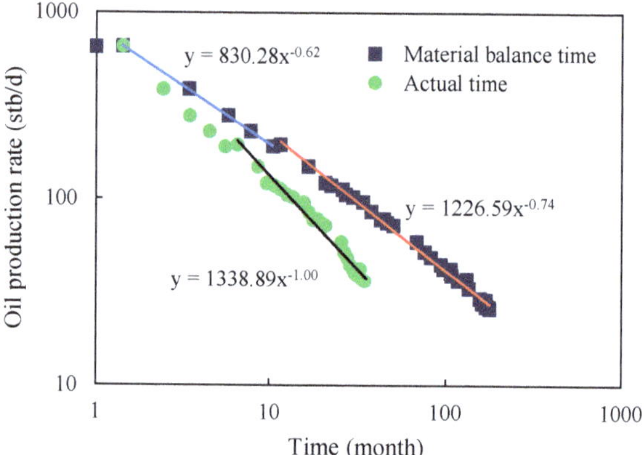

Figure 12. Plot of production rate data for Well #6.

5. Conclusions

In summary, the performances of multifracture horizontal wells (MFHW) in shale oil fields in different regions of the world are mixed. This is believed partially because of the lack of optimization of well completion parameters, especially fracture spacing. Optimization of such spacing is generally recognized as one of the most important steps in enabling economic horizontal wells and requires a lot of attention. A data-driven workflow approach is presented in this study for optimizing fracture spacing of MFHW in shale oil reservoirs. Fracture spacing should be as short as possible unless "interference" occurs, and this "interference" can be found from the pressure transient analysis. If the interference does not happen in initial stages, well productivity can be improved by shortening the fracture spacing. The following conclusions were drawn from this study.

1. This workflow procedure has the advantage of using an analytical well productivity model driven by real production data, making it a practical approach to optimize MFHW in shale oil reservoirs.
2. This workflow procedure employs a closed-form analytical solution and provides a transparent approach to the identification of important fracturing parameters affecting well productivity.
3. This workflow procedure uses transient pressure or production data to identify fracture interference. This offers a reliable and cost-effective means for assessing well production potential in terms of optimizing fracture spacing in the MFHW.
4. Results of a field case study indicated that three wells were drilled and completed with fracture spacing values that were short enough to effectively drain the stimulated reservoir volume (SRV), while the other three wells were drilled and completed with fracture spacings that could be shortened to significantly improve well productivity.

Author Contributions: Data curation, X.Y.; Investigation, B.G.; Methodology, B.G.; Project administration, B.G.; Resources, X.Y.; Validation, X.Y.; Writing—original draft, X.Y.; Writing—review & editing, B.G.

Funding: This research was supported by the U.S. DOE project (Project No. DE-FE0031575).

Conflicts of Interest: The authors declare no conflict of Interest.

Nomenclature

$\overline{p}$	average formation pressure, psia
Δp	pressure difference, psia
B_o	oil formation factor, rb/stb
c	defined by Equation (2)
c_t	total compressibility, psi^{-1}
h_f	fracture height, ft
k_f	fracture permeability, md

k_m matrix permeability, md
L effective lateral length, ft
m slope of inverse of production rate versus $\sqrt{t}$, day$^{1/2}$·stb^{-1}
n_f number of hydraulic fractures
p_e reservoir formation pressure, psia
p_w wellbore pressure, psia
q_o production rate, stb/d
S_f fracture spacing, ft
t_{ehs} time at the end of half slope, days
V_f bulk volume of proppant, ft^3
w average fracture width, inch
x_f hydraulic fracture half-length, ft
η diffusivity, md·psia·cp^{-1}
ϕ porosity, %
μ_o oil viscosity, cp

References

1. Rafiee, M.; Soliman, M.Y.; Pirayesh, E. Hydraulic fracturing design and optimization: A modification to zipper frac. In Proceedings of the SPE Easter Regional Meeting, San Antonio, TX, USA, 8–10 October 2012.
2. Ren, J.; Guo, P. A general analytical method for transient flow rate with the stress-sensitive effect. *J. Hydrol.* **2018**, *565*, 262–275. [CrossRef]
3. Xue, L.; Chen, X.; Wang, L. Pressure transient analysis for fluid flow through horizontal fractures in shallow organic compound reservoir of hydrogen and carbon. *Int. J. Hydrogen Energy* **2019**, *44*, 5245–5253. [CrossRef]
4. Bajwa, A.I.; Blunt, M.J. Early-time 1D analysis of shale-oil and-gas flow. *SPE J.* **2016**, *21*, 1254–1262. [CrossRef]
5. Abbasi, M.; Madani, M.; Sharifi, M.; Kazemi, A. Fluid flow in fractured reservoirs: Exact analytical solution for transient dual porosity model with variable rock matrix block size. *J. Pet. Sci. Eng.* **2018**, *164*, 571–583. [CrossRef]
6. Sesetty, V.; Ghassemi, A. A numerical study of sequential and simultaneous hydraulic fracturing in single and multi-lateral horizontal wells. *J. Pet. Sci. Eng.* **2015**, *132*, 65–76. [CrossRef]
7. Li, J.; Xiao, W.; Hao, G.; Dong, S.; Hua, W.; Li, X. Comparison of different hydraulic fracturing scenarios in horizontal wells using XFEM based on the cohesive zone method. *Energies* **2019**, *12*, 1232. [CrossRef]
8. Du, X.; Nydal, O.J. Flow models and numerical schemes for single/two-phase transient flow in one dimension. *Appl. Math. Model.* **2017**, *42*, 145–160. [CrossRef]
9. Yu, W.; Xu, Y.; Weijermars, R.; Wu, K.; Sepehrnoori, K. Impact of well interference on shale oil production performance: A numerical model for analyzing pressure response of fracture hits with complex cemeteries. In Proceedings of the SPE Hydraulic Fracturing Technology Conference and Exhibition, The Woodlands, TX, USA, 24–26 January 2017.
10. He, Y.; Cheng, S.; Rui, Z.; Qin, J.; Fu, L.; Shi, J.; Wang, Y.; Li, D.; Patil, S.; Yu, H.; Lu, J. An improved rate-transient analysis model of multi-fractured horizontal wells with non-uniform hydraulic fracture properties. *Energies* **2018**, *11*, 393. [CrossRef]
11. Sun, H.; Zhou, D.; Chawathé, A.; Du, M. Quantifying shale oil production mechanisms by integrating a Delaware basin well data from fracturing to production. In Proceedings of the Unconventional Resources Technology Conference, San Antonio, TX, USA, 1–3 August 2016.
12. Orangi, A.; Nagarajan, N.R.; Honarpour, M.M.; Rosenzweig, J.J. Unconventional shale oil and gas-condensate reservoir production, impact of rock, fluid, and hydraulic fractures. In Proceedings of the SPE Hydraulic Fracturing Technology Conference and Exhibition, The Woodlands, TX, USA, 24–26 January 2011.
13. Guo, B.; Yu, X.; Khoshgahdam, M. A simple analytical model for predicting productivity of multifractured horizontal wells. *SPE Reserv. Eval. Eng.* **2009**, *12*, 879–885. [CrossRef]
14. Zhang, C.; Wang, P.; Guo, B.; Song, G. Analytical modeling of productivity of multi-fractured shale gas wells under pseudo-steady flow conditions. *Energy Sci. Eng.* **2018**, *6*, 819–827. [CrossRef]
15. Li, G.; Guo, B.; Li, J.; Wang, M. A mathematical model for predicting long-term productivity of modern multifractured shale gas/oil wells. *SPE Drill. Complet.* **2019**, *34*. [CrossRef]
16. Guo, B.; Liu, X.; Tan, X. *Petroleum Production Engineering*, 2nd ed.; Elsevier: Cambridge, MA, USA, 2017; pp. 432–489. ISBN 978-0-12-809374-0.

17. Potapenko, D.I.; Williams, R.D.; Desroches, J.; Enkababian, P.; Theuveny, B.; Willberg, D.M.; Conort, G. Securing long-term well productivity of horizontal wells through optimization of postfracturing operations. In Proceedings of the SPE Annual Technical Conference and Exhibition, San Antonio, TX, USA, 9–11 October 2017.
18. Feng, F.; Wang, X.; Guo, B.; Ai, C. Mathematical model of fracture complexity indicator in multistage hydraulic fracturing. *J. Nat. Gas Sci. Eng.* **2017**, *38*, 39–49. [CrossRef]
19. Zhang, Q.; Wang, X.; Wang, D.; Zeng, J.; Zeng, F.; Zhang, L. Pressure transient analysis for vertical fractured wells with fishbone fracture patterns. *J. Nat. Gas Sci. Eng.* **2018**, *52*, 187–201. [CrossRef]
20. Li, D.; Zha, W.; Liu, S.; Wang, L.; Lu, D. Pressure transient analysis of low permeability reservoir with pseudo threshold pressure gradient. *J. Pet. Sci. Eng.* **2016**, *147*, 308–316. [CrossRef]
21. Wu, Z.; Cui, C.; Lv, G.; Bing, S.; Cao, G. A multi-linear transient pressure model for multistage fractured horizontal well in tight oil reservoirs with considering threshold pressure gradient and stress sensitivity. *J. Pet. Sci. Eng.* **2019**, *172*, 839–854. [CrossRef]
22. Fekete, F.A.S.T. *Well Test User Manual*; Fekete Associates, Inc.: Calgary, AB, Canada, 2003.
23. E-Production Services, Inc. *PanSystem User Manual*; E-Production Services, Inc.: Edinburgh, UK, 2004.
24. Shan, L.; Guo, B.; Weng, D.; Liu, Z.; Chu, H. Posteriori assessment of fracture propagation in refractured vertical oil wells by pressure transient analysis. *J. Pet. Sci. Eng.* **2018**, *168*, 8–16. [CrossRef]
25. Pang, W.; Wu, Q.; He, Y. Production analysis of one shale gas reservoir in China. In Proceedings of the SPE Annual Technical Conference and Exhibition, Houston, TX, USA, 28–30 September 2015.
26. He, Y.; Cheng, S.; Li, S.; Huang, Y.; Qin, J.; Hu, L.; Yu, H. A semianalytical methodology to diagnose the locations of underperforming hydraulic fractures through pressure-transient analysis in tight gas reservoir. *SPE J.* **2017**, *22*, 924–939. [CrossRef]
27. Cinco, L.H.; Samaniego, V.F. Transient Pressure Analysis for Fractured Wells. *J. Pet. Technol.* **1981**, *33*, 1749–1766. [CrossRef]
28. Gringarten, A.C.; Ramey, H.J.; Raghavan, R. Applied pressure analysis for fractured wells. *J. Pet. Technol.* **1975**, *27*, 887–892. [CrossRef]
29. Cinco, L.H.; Samaniego, V.; Dominguez, A. Transient pressure behavior for a well with a finite-conductivity vertical fracture. *Soc. Pet. Eng. J.* **1978**, *18*, 253–264. [CrossRef]
30. Uzun, I.; Kurtoglu, B.; Kazemi, H. Multiphase rate-transient analysis in unconventional reservoirs: Theory and application. *SPE Reserv. Eval. Eng.* **2016**, *19*, 553–566. [CrossRef]
31. Yang, C.; Sharma, V.K.; Datta-Gupta, A.; King, M.J. Novel approach for production transient analysis of shale reservoirs using the drainage volume derivative. *J. Pet. Sci. Eng.* **2017**, *159*, 8–24. [CrossRef]
32. Marsden, J.; Kostyleva, I.; Fassihi, M.R.; Gringarten, A.C. A conceptual shale gas model validated by pressure and rate data from the Haynesville shale. In Proceedings of the SPE Annual Technical Conference and Exhibition, San Antonio, TX, USA, 9–11 October 2017.
33. Ambrose, R.J.; Clarkson, C.R.; Youngblood, J.E.; Adams, R.; Nguyen, P.D.; Nobakht, M.; Biseda, B. Life-cycle decline curve estimation for tight/shale reservoirs. In Proceedings of the SPE Hydraulic Fracturing Technology Conference, The Woodlands, TX, USA, 24–26 January 2011. [CrossRef]
34. Wattenbarger, R.A.; El-Banbi, A.H.; Villegas, M.E.; Maggard, J.B. Production analysis of linear flow into fractured tight gas wells. In Proceedings of the SPE Rocky Mountain Regional/Low-Permeability Reservoirs Symposium, Denver, CO, USA, 5–8 April 1998. [CrossRef]

© 2019 by the authors. Licensee MDPI, Basel, Switzerland. This article is an open access article distributed under the terms and conditions of the Creative Commons Attribution (CC BY) license (http://creativecommons.org/licenses/by/4.0/).

Article

Impact of Shale Anisotropy on Seismic Wavefield

Han Li [1,2,3,4], Xiwu Liu [1,2], Xu Chang [3,4,*], Ruyue Wu [3,4] and Jiong Liu [1,2]

1 State Energy Center for Shale Oil Research and Development, Sinopec, Beijing 100083, China; lihan@mail.iggcas.ac.cn (H.L.); liuxw.syky@sinopec.com (X.L.); liujiong.syky@sinopec.com (J.L.)
2 State Key Laboratory of Shale Oil and Gas Enrichment Mechanisms and Effective Development, Beijing 100083, China
3 Key Laboratory of Shale Gas and Geoengineering, Institute of Geology and Geophysics, Chinese Academy of Sciences, Beijing 100029, China; wuruyue@mail.iggcas.ac.cn
4 Innovation Academy for Earth Sciences, Chinese Academy of Sciences, Beijing 100029, China
* Correspondence: changxu@mail.iggcas.ac.cn; Tel.: +86-10-8299-8281

Received: 21 October 2019; Accepted: 18 November 2019; Published: 20 November 2019

Abstract: During unconventional resources exploration, ignoring shale anisotropy may lead to wrong seismic interpretations, thus affecting the accuracy and credibility of sweet spots prediction and reservoir characterization. In order to investigate the impact of shale anisotropy on the seismic wavefield, we propose a quantitative evaluation method by calculating the waveforms' amplitude and phase deviations. Based on the 3D elastic wave equation and the staggered-grid finite-difference method, the forward modeling theory with the consideration of shale anisotropy is established. Then, we use the envelope misfit (EM) and phase misfit (PM) parameters to illustrate the differences in waveforms' amplitude and phase morphology, which are caused by anisotropy. Lastly, by comparing the waveforms of the models with/without anisotropy and calculating their EM and PM values, a practical and quantitative evaluation method is constructed. We used synthetic models of different complexity and oilfield models to validate the proposed method. Through the research, we also gained some new insights about the anisotropy's effects. For a certain medium model, the impact of shale anisotropy on seismic wavefield is complicated and needs specific analysis. The proposed method provides a useful and quantitative tool for the evaluation of shale anisotropy's impact.

Keywords: shale reservoir; anisotropy; seismic wavefield; quantitative evaluation

1. Introduction

At present, more and more attention is paid on the research of unconventional resources exploration [1,2]. Unconventional resources, especially shale oil and gas, are realistic backup resources to meet the demand for fossil energy [3]. The shale oil and gas deposits are mainly stored in tight shale rocks or mud rocks. Shale is a sedimentary rock which is composed of fine debris, clay, and organic matter with a particle size of less than 0.0039 mm [4]. Rock physics research shows that shale has strong anisotropic characteristics (up to 70%) [5,6] and most shale reservoirs currently developed have distinct seismic anisotropy [7–12].

Due to the strong seismic anisotropy of shales, seismic waves propagating through it will be influenced and changed. Therefore, if anisotropy is ignored, interpretations may lead to wrong models. Carcione [13] analyzed the influence of formation anisotropy on the amplitude of reflected seismic phases (PP and PS) and found that anisotropy of shale must be considered in AVO (amplitude variation with offset) analysis. Research in borehole microseismic monitoring for the hydraulic fracturing [14,15] concluded that failure to account for the shale anisotropy results in significant microseismic event location errors, and methods must be developed to circumvent the effects of anisotropy or to determine the anisotropy parameters. Tsvankin [16] reviewed state-of-the-art in modeling, processing, and inversion of seismic data for anisotropy media and concluded that the anisotropy has a strong influence on seismic data, especially for the shear and mode-converted wavefields. Through the forward modeling research based on VTI (vertical transversely isotropic) acoustic wave equation, Sun [17] found that anisotropy has large influence on the phases and amplitudes of seismic waves and must be considered when processing seismic data. Meléndez-Martínez and Schmitt [18] illustrated that the problems introduced by attempting to use Poisson's ratio to estimate fracture gradient are even further from reality once anisotropy is included. Malehmir and Schmitt [19,20] developed an algorithm to solve reflectivity, transmissivity, velocity, and particle polarization in the case of elastic anisotropy, and found that the tilt of the symmetry planes of anisotropic geological formations will influence observations in AVAz (amplitude versus azimuth) field studies, and knowledge of this tilt becomes essential to fully understand the properties of the subsurface.

In recent years, there has been growing interest in shale anisotropy. However, only few studies have been performed on shale anisotropy's effects on 3D elastic seismic wavefield and quantitative evaluation method. Compared with the acoustic wave, an elastic wave formulation is likely to be a more realistic representation of the behavior of real seismic waves. The quantitative evaluation of anisotropy can identify the observation area where the detectors are less affected by the shale anisotropy. Then in these areas, less affected by anisotropic nature of the substrate, we can apply traditional seismic interpretation and processing method to the selected seismic data, thus improving the accuracy and reliability of sweet spots prediction and reservoir characterization. In this paper, based on the 3D elastic wave equation and the staggered-grid finite-difference method [21,22], the seismic wave forward modeling theory with shale anisotropy was established. Then, by calculating the deviations in waveforms' amplitude and phase which are caused by anisotropy, we developed a quantitative evaluation method. Lastly, we set several models according to the shale reservoir characteristics or the actual oilfield in China, added different levels of anisotropy to them, and applied the proposed method to analyze the shale anisotropy's effects on their seismic wavefield.

2. Methodology

2.1. Forward Modeling for 3D TTI Model

Considering the shale's strong anisotropy characteristics, the assumption of the isotropic medium is inappropriate, and the adoption of transversely isotropic media is currently recognized as a more reasonable description of shale media [23,24]. In our study, the calculation of the seismic wavefield is based on the elastic wave equation:

$$\rho \frac{\partial^2 u_i}{\partial t^2} = c_{ijkl} \frac{\partial^2 u_k}{\partial x_j \partial x_l} + \rho F_i, \tag{1}$$

where u represents the particle's displacement, ρ and F are the medium's density and the external force, respectively, c_{ijkl} is the stiffness matrix in Hooke's law and represents the relationship between stress and strain. Considering the symmetry of stress and elastic coefficient tensors, and assuming that the medium is vertical transversely isotropic (VTI), the stiffness matrix c_{ijkl} can be simplified as:

$$\begin{bmatrix} c_{11} & c_{12} & c_{13} & 0 & 0 & 0 \\ c_{12} & c_{11} & c_{13} & 0 & 0 & 0 \\ c_{13} & c_{13} & c_{33} & 0 & 0 & 0 \\ 0 & 0 & 0 & c_{44} & 0 & 0 \\ 0 & 0 & 0 & 0 & c_{44} & 0 \\ 0 & 0 & 0 & 0 & 0 & c_{66} \end{bmatrix}, \tag{2}$$

using the weak anisotropy hypothesis proposed by Thomsen [25], the elastic parameters in (2) can be transformed to:

$$\begin{cases} \varepsilon = \frac{c_{11}-c_{33}}{2c_{33}}, \ \gamma = \frac{c_{66}-c_{44}}{2c_{44}}, \\ v_{p_z} = \sqrt{\frac{c_{33}}{\rho}}, \quad v_{s_z} = \sqrt{\frac{c_{44}}{\rho}}, \\ \delta = \frac{(c_{13}+c_{44})^2-(c_{33}-c_{44})^2}{2c_{33}(c_{33}-c_{44})}, \end{cases} \quad \begin{cases} c_{11} = \rho(1+2\varepsilon)v_{p_z}, \\ c_{33} = \rho v_{p_z}^2, \\ c_{44} = \rho v_{s_z}^2, \\ c_{12} = \rho v_{p_z}^2[1+2\varepsilon-2(1-f')(1+2\gamma)], \\ c_{13} = \rho v_{p_z}^2 \sqrt{f'(f'+2\delta)} - \rho v_{s_z}^2, \\ c_{66} = \rho(1+2\gamma)v_{s_z}^2 = 0.5(c_{11}-c_{12}), \end{cases} \quad (3)$$

where $f' = 1 - v_{s_z}^2/v_{p_z}^2$, ε, and δ represent the difference of P-wave velocity in the direction of vertical, horizontal, and 45°, i.e., the measurement of P-wave anisotropy, δ represent the anisotropic strength of S-wave, v_{p_z} and v_{s_z} are the velocity of P- and S-wave along the vertical (depth) direction, respectively. From Equations (2) and (3), we can see that for the VTI medium, there are six stiffness parameters and five of them are independent with each other.

Equations (1)–(3) establish the elastic wave forward modeling theory of VTI medium (vertical transversely isotropic, i.e., medium with vertical symmetry axis). For TTI medium (tilted transversely isotropic, i.e., medium with tilted symmetry axis), we can make use of its spatial inclination and perform a coordinate transformation to calculate the seismic wavefield.

In this paper, we use the staggered-grid finite-difference method [21,22,26–29] to calculate the elastic wave equation. Details of this method are shown in Appendix A. The finite-difference (FD) method is a crucial numerical tool in the modeling of earthquake ground motion, and the staggered-grid FD algorithm is one of the most popular FD schemes and has been proved to be flexible and relatively accurate in the analysis of wave propagation problems [27,29,30]. The staggered-grid FD algorithm computes the pressure at a set of spatial points, and the velocity at another set of spatial points [31]. One of the attractive features of the staggered-grid approach is the velocities are updated independently from the stresses, which allows for a very efficient and concise implementation scheme [27].

2.2. *Quantitative Evaluation of the Shale Anisotropy's Impact*

In order to analyze the shale anisotropy's impact on the seismic wavefield, the comparison between seismic data with and without anisotropy is needed. We use the envelope misfit EM and the phase misfit PM to perform the quantitative evaluation. These two parameters are put forward by Kristek [30] to evaluate both the difference in the amplitude and phase between two signals. Let $S_{REF}(t)$ be the reference data and $S(t)$ be the data to be compared, the envelope misfit EM is

$$EM = \frac{\sqrt{\sum_m \left[\left|\hat{S}_{REF}(t_m)\right| - \left|\hat{S}(t_m)\right|\right]^2}}{\sqrt{\sum_m \left|\hat{S}_{REF}(t_m)\right|^2}} \quad (4)$$

and the phase misfit PM is

$$PM = \frac{\sqrt{\sum_m \left[\left|\hat{S}_{REF}(t_m)\right| \cdot Arg\left(\hat{S}_{REF}(t_m)/\hat{S}(t_m)\right)\right]^2}}{\pi \cdot \sqrt{\sum_m \left|\hat{S}_{REF}(t_m)\right|^2}} \quad (5)$$

where $\hat{S}_{REF}(t)$ and $\hat{S}(t)$ are the analytical signals of $S_{REF}(t)$ and $S(t)$, respectively (an analytical signal is a complex-valued function that has no negative frequency components, its real and imaginary parts are real-valued functions related to each other by the Hilbert transform [32]). $|z|$ is the modulus of a complex number z and $Arg(z)$ is the principal value of z's argument. From Equations (4) and (5), it can be seen that the envelope misfit EM and the phase misfit PM are the relative average errors of two waveforms. If the value of EM is 1, it means that the average amplitude difference between the two signals is doubled. For PM, if its value is 1, the polarities of all seismic phases in two signals are completely opposite.

In this paper, we use the following steps (Figure 1) to quantitatively evaluate the impact of shale anisotropy on seismic wavefield:

1. Set the parametric medium model M_0 (without anisotropy) according to the geological characteristics of shale reservoirs or the actual oilfield models. Add three sets of different anisotropic parameters to the shale layer of model M_0 and establish three different medium models with shale anisotropy, i.e., M_E ($\varepsilon = 0.25$, $\delta = 0$), M_D ($\varepsilon = 0$, $\delta = 0.25$), and M_{ED} ($\varepsilon = 0.25$, $\delta = 0.25$). The parameters of model M_{ED} are reasonably geologically chosen from several shale anisotropy studies in China ([9,17]), while the other two models (M_E and M_D) are built based on the variable-controlling approach, for the purpose of exploring the impact of different anisotropy parameters;
2. Use the forward modeling method to calculate the elastic wavefields of four medium models (S_{REF} for model M_0, S_E for model M_E, S_D for model M_D, and S_{ED} for model M_{ED});
3. Compare the seismic wavefields with each other and calculate the envelope misfit EM and the phase misfit PM of the wavefield $S_E/S_D/S_{ED}$ from S_{REF}. Evaluate the impact of shale anisotropy on elastic seismic wave response.

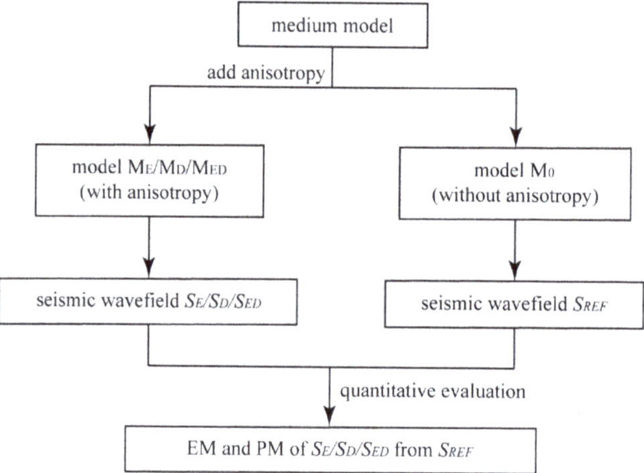

Figure 1. The flow chart of the proposed method for the analysis of anisotropy effects. The procedure of this method starts with adding anisotropy to the medium model, followed by the seismic wavefield forward modeling and quantitative evaluation.

3. Evaluation of Simulation Models

In this section, we set two synthetic models of different complexity and used our method to evaluate the impact of anisotropy on their seismic wavefield. Based on the geological sedimentary characteristics of actual shale strata, we set two models, i.e., the horizontal layered VTI model and the curved layered TTI model.

3.1. Horizontal Layered VTI Model

3.1.1. Parameters and Wavefield Data

The horizontal layered VTI model, seismic source, and observation system are shown in Figure 2. This model has three layers and the shale reservoir is located in the middle. Details of each layer's parameters are shown in Table 1. An explosion source is situated in the center of the surface (1000, 1000, 0 m) with the Ricker wavelet time function (20 Hz). There are two crossing survey lines distributed along the X direction and the Y direction, respectively, each with 201 geophones and 10 m spacing. The grid number is 201 × 201 × 201 (X/Y/Z) with the spatial spacing 10 m and the time sampling interval is 0.5 ms. According to the steps in Section 2.2 (Figure 1), the horizontal layered VTI model in Figure 2 is set as model M_0 (without anisotropy). We added three groups of anisotropic parameters to the shale layer and built up three different anisotropic models, i.e., M_E ($\varepsilon = 0.25$, $\delta = 0$), M_D ($\varepsilon = 0$, $\delta = 0.25$), and M_{ED} ($\varepsilon = 0.25$, $\delta = 0.25$).

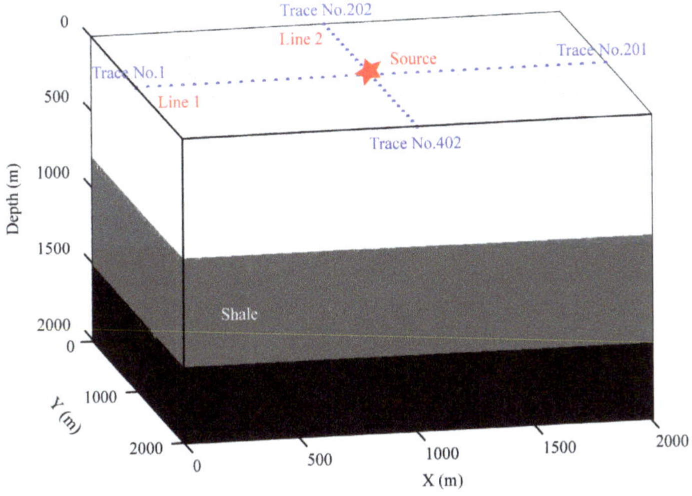

Figure 2. Multi-layered VTI model, source, and observation system. The red star is the source; the blue dots are the detectors distributed on two survey lines (402 geophones in total, with 10 m spacing).

Table 1. Media parameters for multi-layered VTI model.

Layer	Depth (m)	V_P (m/s)	V_S (m/s)	Density (g/cm^3)
1	0~800	3000	1700	2.0
2	800~1500	3500	2000	2.3
3	1500~2000	4000	2300	2.4

For models $M_0/M_E/M_D/M_{ED}$, the seismic waveform records of 402 geophones are respectively synthesized (S_{REF} / $S_E/S_D/S_{ED}$). The three-component waveforms of model M_0, namely S_{REF}, is shown in Figure 3a–c. Comparing the X-, Y-, and Z-component waveforms of each detector, we can see that the X- and Y-component records of the two survey lines (line 1: No.1~201 and line 2: No.202~402) are different, while the Z-component records are entirely consistent. The reason for this is that the horizontal coordinates of the two survey lines are different (line 1 is arranged along the X direction, and line 2 is along the Y direction). Therefore, for an explosive seismic source, the received X- and Y-waveforms are different, while the Z-component records of two lines are the same, because the symmetry axis of the medium is along the vertical direction (VTI medium).

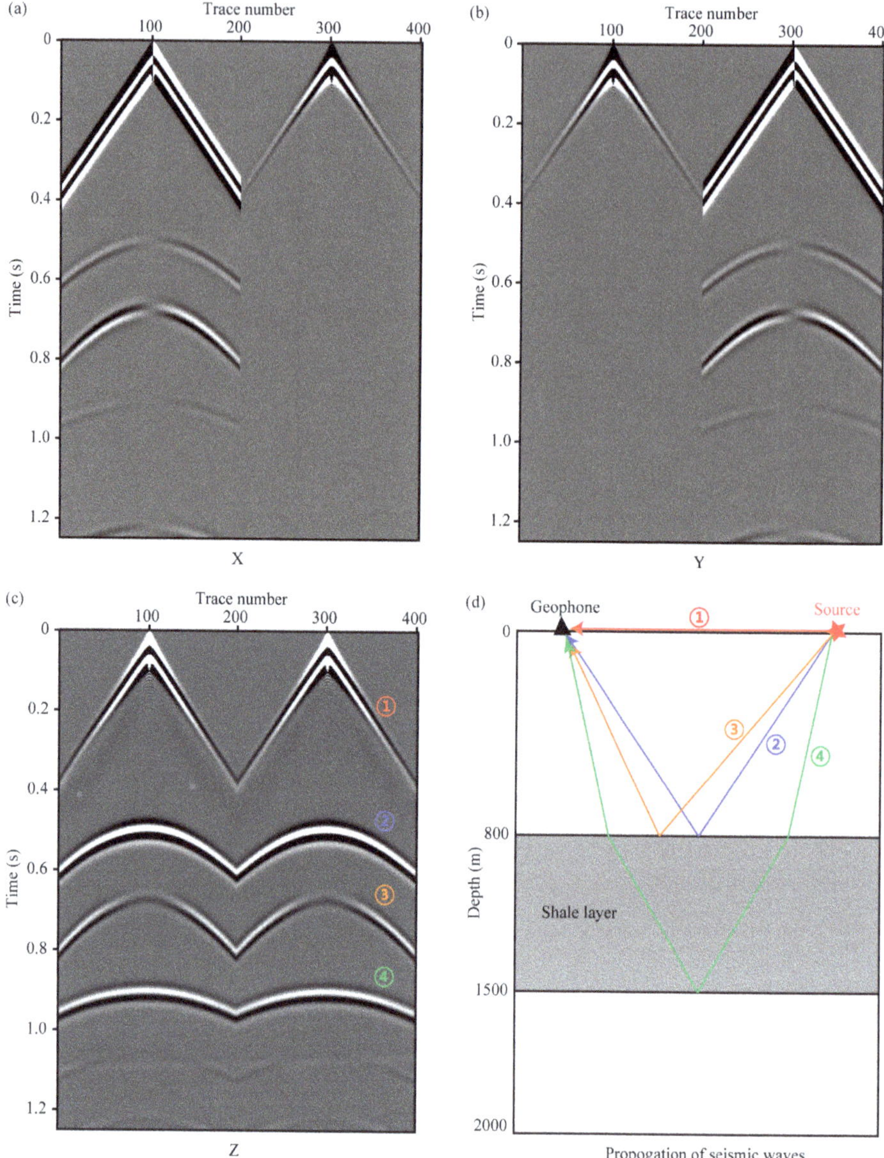

Figure 3. Three-component seismic wavefield of the M_0 model and corresponding wave propagation diagrams of four seismic phases. (**a**)/(**b**)/(**c**) are the geophones' X/Y/Z waveform records, respectively, (**d**) indicates the seismic wave propagation paths corresponding to each phase, with ① direct P wave, ② reflected P-P wave, ③ reflected P-SV wave, and ④ P-P-P-P wave.

Considering that surface seismic exploration usually uses a single-component (generally Z-component) detector, our study mainly focuses on analyzing the influence of shale anisotropy on the Z-component waveforms. It can be seen from Figure 3c that the Z-component wavefield has four strong phases. Based on the analysis in Figure 3d, these four phases respectively correspond to direct P wave, reflected P-P wave, reflected P-SV wave, and P-P-P-P wave.

3.1.2. Effect of Anisotropy on Different Seismic Phases

Seismic waves with different phases (direct P/S, reflected, or transmitted wave) propagate in different paths, so the impact of shale anisotropy on them may also be different. Figure 4 shows the comparison of waveforms from model M_{ED} ($\varepsilon = 0.25$, $\delta = 0.25$) and M_0 ($\varepsilon = 0$, $\delta = 0$).

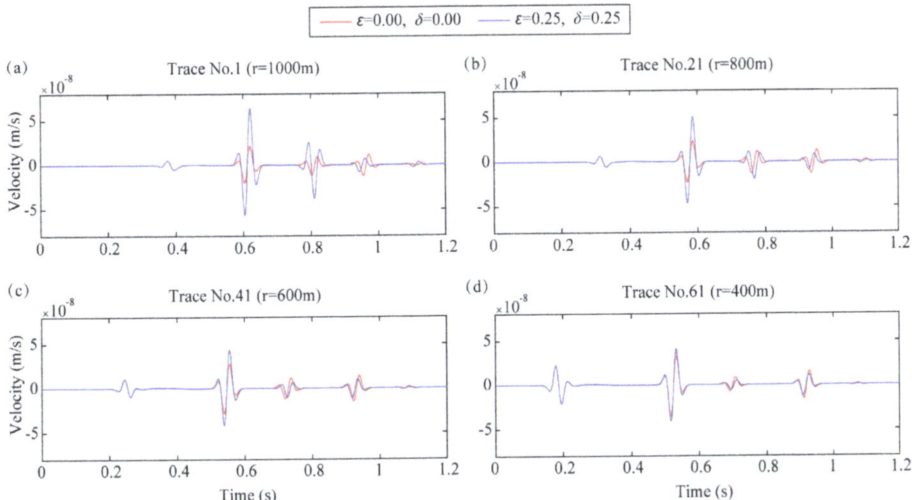

Figure 4. Comparison of Z-component waveforms from model M_{ED} and (with/without anisotropy, VTI). (**a**)/(**b**)/(**c**)/(**d**) are waveforms of geophones No.1/21/41/61 (with epicentral distance 1000/800/600/400 m), respectively. Red indicates seismic waveforms of M_0 (without anisotropy); blue indicates seismic waveforms from M_{ED} (with anisotropy).

From Figure 4, we can find that for each detector:

1. Direct P wave. The first seismic phases (direct P wave) of two models are exactly the same. From its propagation path (Figure 3d), we can see that the direct P wave traveled from the seismic source to the detector through the surface, so it is not affected by the shale anisotropy.
2. Transmitted and reflected P-P-P-P wave. The fourth seismic phases (P-P-P-P wave, its propagation path is shown in Figure 3d) are totally different. When adding anisotropy to the shale layer, the velocities of P-waves along different directions propagating in it are changed and not the same. So this seismic wave's arrival time, amplitude and phase are all influenced by shale anisotropy.
3. Reflected P-P wave and P-SV wave. The arrival time of the second and third phases (reflected P-P and P-SV waves, their paths are shown in Figure 3d) is the same, while their amplitudes are different. The reasons for this are probably because:

   - Arrival time. These two waves are both reflected at the interface of the first layer and the shale layer. The first layer is isotropic, so we can use the Snell's law to analyze the reflection:

$$\frac{v_{P1}}{\sin\theta_1} = \frac{v_{S1}}{\sin\phi_1}, \tag{6}$$

   where v_{P1} is the velocity of incident P-wave (same as the reflected P-wave's velocity) and v_{S1} is the reflected SV-wave's velocity at the interface, respectively, θ_1 and ϕ_1 are the corresponding angles (Figure 5a). When we added anisotropy to the shale layer, velocities of P and S wave in the first layer are not changed, so v_{P1} and v_{S1} remain unchanged. From Equation (6), we can find that θ_1, θ_2, and the propagation paths are not influenced by anisotropy, so is the arrival time.

- Amplitude. The amplitude of reflected P and SV waves can be evaluated using the anisotropic reflectivity and transmissivity calculator code by Malehmir and Schmitt [19]. According to the observation system and source location (Figure 2), the maximum incident angle θ_1 of P wave between the first and second (shale) layer is about 32°. Then, we set θ_1 from 0 to 35° and calculate the reflection coefficients of reflected P and SV waves. The differences of reflection coefficients with and without anisotropy are shown in Figure 5b. We can find that when adding anisotropy to the VTI model (Figure 2), the reflection coefficients of reflected P and SV waves are both changed. Considering that the propagation paths of these two waves are unchanged, so the Z-component amplitudes of reflected P-P and P-SV waves are influenced by the shale anisotropy.

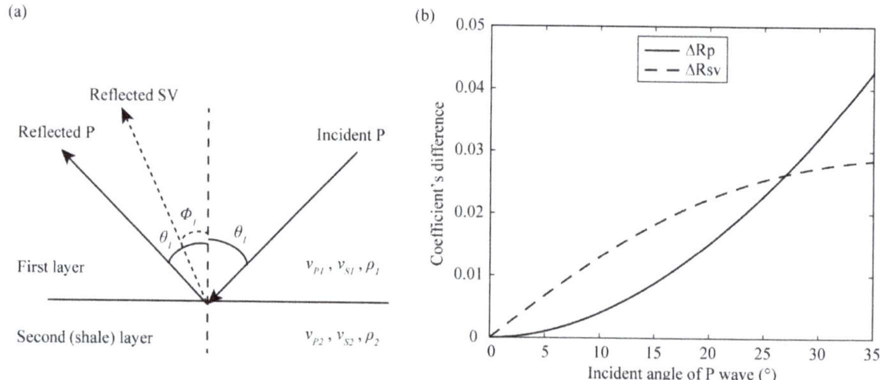

Figure 5. Schematic diagram and coefficients' differences of reflected waves at the interface. (**a**) Reflection at the interface between first and second (shale) layer in the VTI model (Figure 2). θ_1 and ϕ_1 are the angles of P and SV waves, respectively. (**b**) Reflection coefficients' differences of models with and without anisotropy (M_{ED} and M_0). Solid and dash line indicates reflected P and SV waves, respectively.

3.1.3. Amplitude and Phase Misfit by Anisotropy

Using the forward modeling method in Section 2.1, we synthesized the seismic waveforms of the four medium models $M_0/M_E/M_D/M_{ED}$ (ε/δ are 0/0, 0.25/0, 0/0.25, and 0.25/0.25). Based on Equations (4) and (5), we calculated the wavefields' amplitude deviation EM and phase deviation PM of the anisotropic model $M_E/M_D/M_{ED}$ from the model M_0 (without anisotropy). The results are shown in Figure 6. Here, the normalized epicentral distance (i.e., epicentral distance divided by wavelength) is used. From Figure 6, we can see the following points:

- The amplitude deviation EM and phase deviation PM of the two survey lines in Z-component are significantly larger than their X- and Y-components, which indicates that anisotropy has a more significant influence on the vertical component. It is speculated that the reason may be that the propagation distance of the seismic wave in the vertical direction is larger than that in the horizontal direction and the medium model's symmetric axis is along the vertical direction.
- The maximum EM of the two survey lines in Z-component is greater than 1, which indicates that the waveform's amplitude will be significantly affected by the shale anisotropy. In actual seismic exploration, the amplitude is of great importance to the inversion of reservoir parameters, so ignoring anisotropy may lead to errors in reservoir characterization.
- The Z-component phase deviation PM of the two survey lines reaches 0.4. As is mentioned above, if PM is 1, the polarities of the two signals are completely opposite. So this means that the

waveforms' phase morphology is also largely changed due to anisotropy. The inaccurate phase may lead to the low resolution of migration imaging results, which also affects the processing and interpretation of actual exploration seismic data.

- The amplitude deviation EM and phase deviation PM of X/Y/Z components on two sides of the source are symmetrical, which is caused by the symmetry axis's verticality of the VTI model. Moreover, the EM all increases with the increase of epicentral distance (offset), while the PM increases first and then decreases. This may because the maximum value of the difference in phase is the odd multiples of π. Therefore, if the phase's difference gradually increases from 0 to 2π, the difference in waveform's phase will become bigger first and then smaller, and the calculated PM will also increase first and then decrease correspondingly.
- In most of the results, the deviations of the model M_{ED} ($\varepsilon = 0.25$, $\delta = 0.25$) are relatively larger than the other two models M_E ($\varepsilon = 0.25$, $\delta = 0$) and M_D ($\varepsilon = 0$, $\delta = 0.25$).

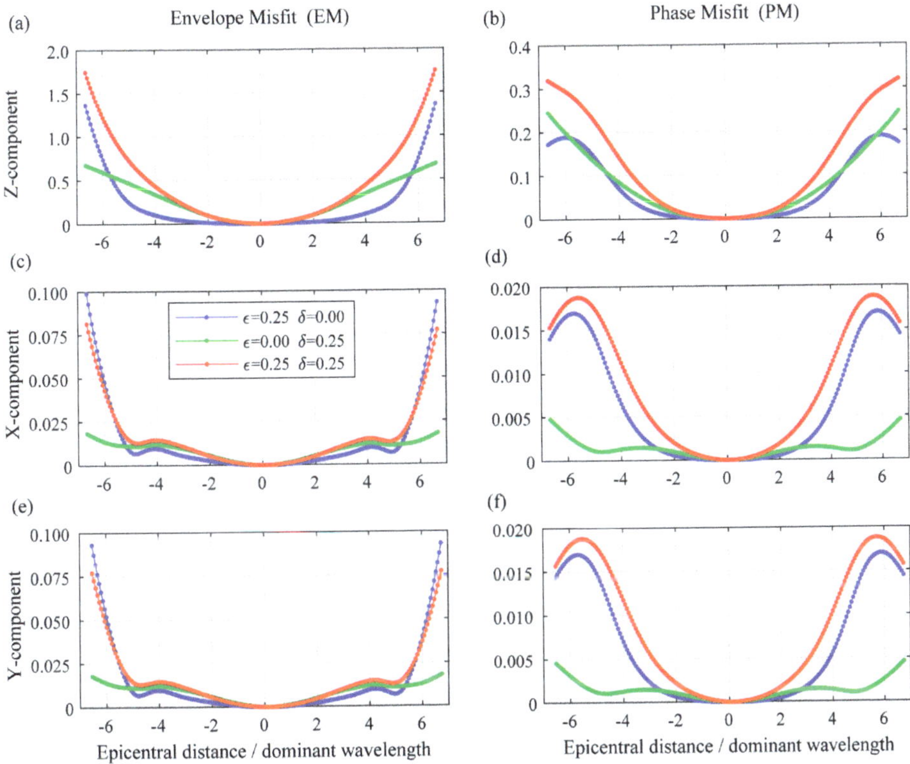

Figure 6. Envelope and phase misfit of detectors' three-component waveforms (VTI model). The blue/green/red lines respectively represent the waveforms' deviation curve of M_E ($\varepsilon = 0.25$, $\delta = 0$)/M_D ($\varepsilon = 0$, $\delta = 0.25$)/M_{ED} ($\varepsilon = 0.25$, $\delta = 0.25$) with respect to M_0 ($\varepsilon = 0$, $\delta = 0$), and the horizontal axis is the normalized epicentral distance (epicentral distance divided by wavelength) of a detector from the seismic source, the sign indicates that the detectors are located on two sides of the source (positive along X/Y axis direction). (**a**,**b**) are the envelope and phase misfit recorded by Z-component of line 1, (**c**,**d**) are the X-component's deviations of line 1, (**e**,**f**) are the deviations recorded by Y-component of line 2.

3.2. Curved Layered TTI Model

In this section, we analyzed shale anisotropy's impact on the seismic wavefield of a curved multi-layer TTI model [28]. The geometry of the medium model, source, and observation system is shown in Figure 7. Parameters for the wavefields' forward modeling are the same as Section 3.1. Details of each layer's parameters are listed in Table 2 and the shale reservoir is located in the middle. The dip angle of TTI media's symmetry axis is 40° and the azimuth angle is 0°. Similar to Section 3.1, when anisotropic parameters are added to the shale layer, four models are respectively set: M_0 ($\varepsilon = 0$, $\delta = 0$), M_E ($\varepsilon = 0.25$, $\delta = 0$), M_D ($\varepsilon = 0$, $\delta = 0.25$), and M_{ED} ($\varepsilon = 0.25$, $\delta = 0.25$).

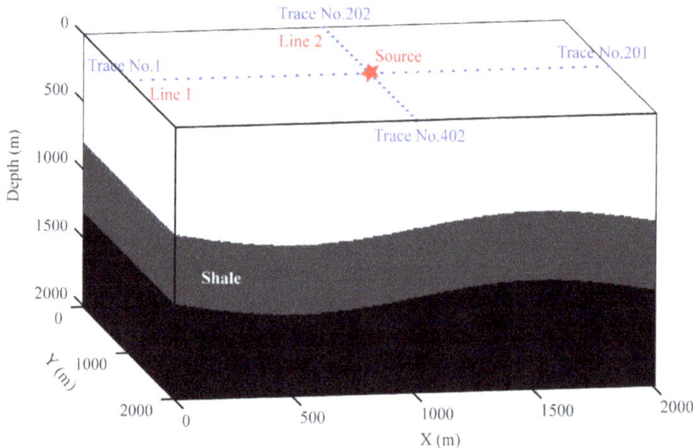

Figure 7. Curved layered TTI model, source, and observation system. The red star is the source; the blue dots are the detectors distributed on two survey lines (402 detectors in total, with 10 m spacing).

Table 2. Media parameters for multi-layered TTI model.

Layer	Depth (m)	Thickness (m)	V_P (m/s)	V_S (m/s)	Density (g/cm^3)
1	0	700–900	3000	1700	2.0
2	800	500	3500	2000	2.3
3	1500	600–800	4000	2300	2.4

The seismic waveform records of 402 geophones are respectively forward synthesized for four models M_0, M_E, M_D, and M_{ED}. The comparison of Z-component wavefields with and without anisotropy in the first survey line (geophone No.1–No.201) is shown in Figure 8, and the waveforms of eight different epicentral distance detectors (−1000/−600/−200/+200/+400/+1000 m, the positive and the negative sign indicate that the detectors are on left and right sides of the seismic source, respectively) are shown in Figure 9. From Figures 8 and 9, it can be seen that before and after anisotropy is added, except the direct P wave is not affected, the subsequent phases are all affected by different degrees. When adding shale anisotropy, the amplitude of reflected P-P wave (the second phase) is increased. The arrival time, phase, and amplitude of the other phases (P-SV, P-P-P-P, multiple waves, etc.) are all changed due to the shale anisotropy.

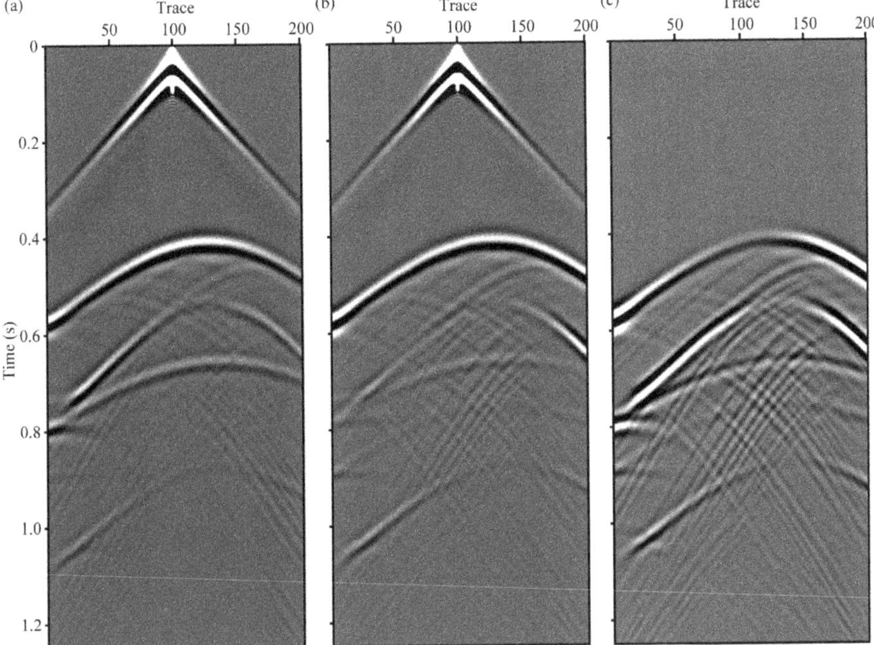

Figure 8. Comparison of Z-component waveforms in line 1 with and without the shale anisotropy (TTI model). (**a**) is the Z-component wavefield of model M_0 (without anisotropy), (**b**) is the Z-component wavefield of model M_{ED} (with anisotropy), and (**c**) is the residual wavefield for (**a**) and (**b**).

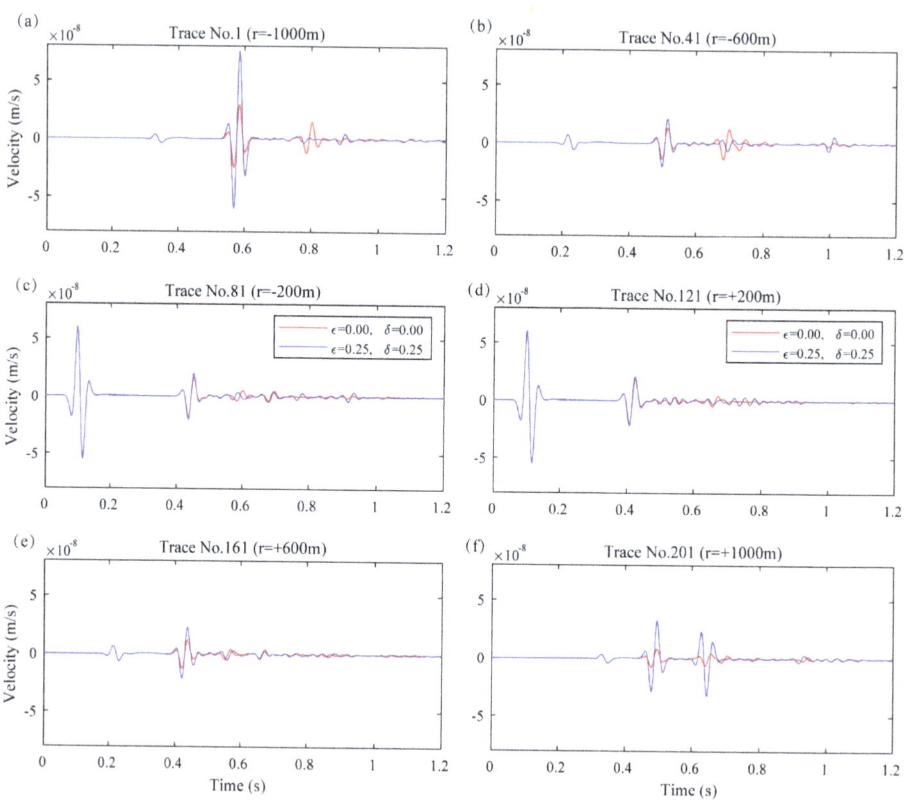

Figure 9. Comparison of Z-component waveforms from model M_{ED} and M_0 (with/without anisotropy, TTI). (**a**)–(**f**) are waveform records of geophones No.1/41/81/121/161/201 (with epicentral distance −1000/−600/−200/+200/+600/+1000 m, the sign indicates that the detectors are located on two sides of the source, positive along X-axis direction). Red indicates synthetic waveforms from the model M_0 (without anisotropy); blue indicates seismic waveforms from the model M_{ED} (with shale anisotropy).

Similar as Section 3.1, using Equations (4) and (5), we calculated the seismic wavefields' amplitude deviation (envelope misfit EM) and phase deviation (phase misfit PM) of the anisotropic model M_E/M_D/M_{ED} (ε/δ are 0.25/0, 0/0.25, 0.25/0.25) from the model M_0 (without anisotropy). The results are shown in Figure 10. Here, the deviation results of Z- and X-component are from line 1, while the results of Y-component are from line 2. From Figure 10, we can find that:

- Same as the VTI model, the EM and PM of Z-component are still larger than that of X- and Y-components. The maximum EM is still greater than one, and the maximum PM is about 0.5 (relatively large; if PM is 1, the polarities of all phases of the two signals are entirely opposite).
- Unlike the VTI model, with the increase of epicentral distance, the variation trend of EM and PM are complicated. Moreover, the EM and PM of Y-component are symmetrical, while the X- and Z-components are not. The reasons are probably because the shape and structure of TTI model are complex, and its symmetry axis is not along the vertical direction. These phenomena prove that the impact of shale anisotropy relies heavily on the model.
- In most of the results, the deviations of models M_{ED}, M_E, and M_D from M_0 are close to each other. This indicates that the impact of different anisotropic parameters on the wavefield is complicated in the curved TTI model, and the influence strength of each parameter cannot be determined simply as the horizontal layered VTI model.

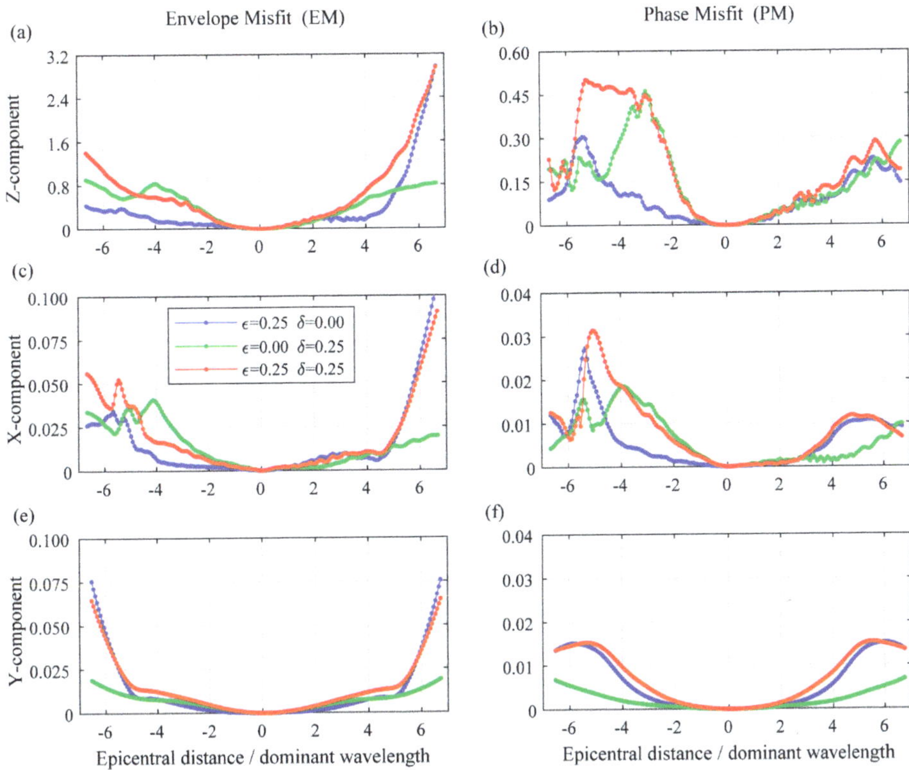

Figure 10. Envelope misfit and phase misfit of detectors' three-component waveforms (TTI model). The meanings of each line and axis are similar to Figure 6.

4. Evaluation of JY Depression Model

Through the analysis of the simulation VTI and TTI model, we find that if the medium model is complex (shape is distorted and the symmetry axis is inclined), the influence of shale anisotropy on wavefield is complicated and requires specific analysis. In this section, we try to apply the established evaluation method (Figure 1) to quantitatively analyze the anisotropic influence of an actual shale model (Jiyang depression model) in eastern China from China Petrochemical Corporation (Sinopec).

Figure 11 shows the geometry and velocity parameters of the 3D JY (Jiyang) depression model, and the location of the seismic source and two survey lines (with 402 geophones). When setting the JY depression model, we did some compression along the X direction and conducted continuation modeling along the Y direction. The shale layer (red) is located in the middle and we add four groups of different anisotropic parameters to it, thus set four models, i.e., M_0 ($\varepsilon = 0$, $\delta = 0$), M_E ($\varepsilon = 0.25$, $\delta = 0$), M_D ($\varepsilon = 0$, $\delta = 0.25$), and M_{ED} ($\varepsilon = 0.25$, $\delta = 0.25$). The grid number of the model is $201 \times 201 \times 201$ with spatial size $10 \times 10 \times 10$ m and the time sampling interval is 0.5 ms. The explosive source is located in the center of the surface (1000, 1000, 0 m) and the source time function is Ricker wavelet (20 Hz).

For four models ($M_0/M_E/M_D/M_{ED}$, with different shale anisotropy), the seismic waveform records of each detector are forward simulated, respectively. Figure 12 shows the comparison of the Z-component wavefield of line 1 (trace numbers No.1–No.201, Figure 11) with and without the shale anisotropy. Figure 13 illustrates the comparison of Z-component waveforms of eight detectors (with different epicentral distances).

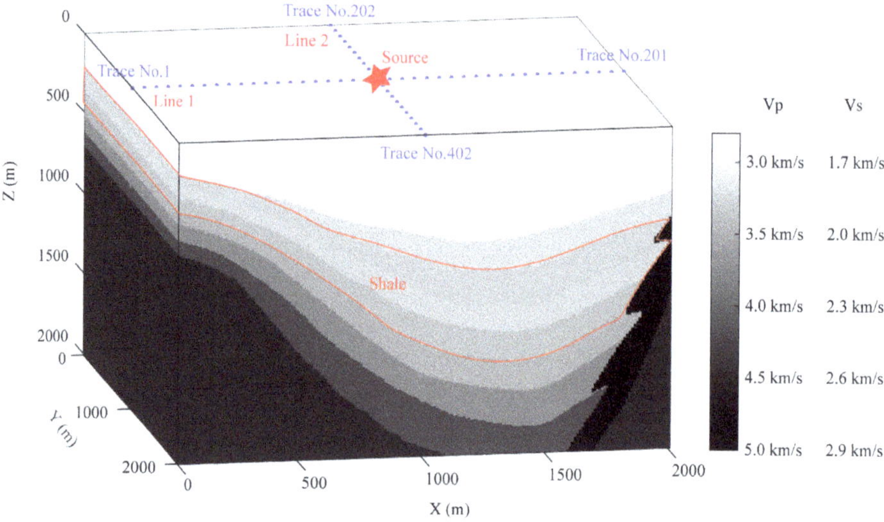

Figure 11. JY depression model, source, and observation system. The red star is the source; the blue dots are the detectors distributed on two survey lines (402 detectors in total, with 10 m spacing).

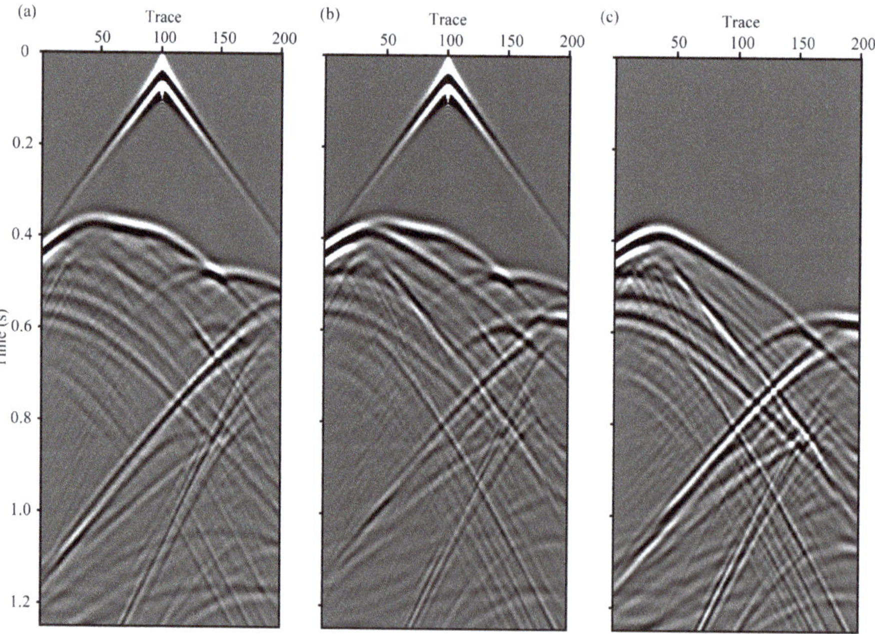

Figure 12. Comparison of Z-component waveforms in line 1 with and without the shale anisotropy (JY model). (**a**) is the Z-component wavefield of model M_0 (without anisotropy), (**b**) is the Z-component wavefield of model M_{ED} (with anisotropy), and (**c**) is the residual wavefield for (**a**) and (**b**).

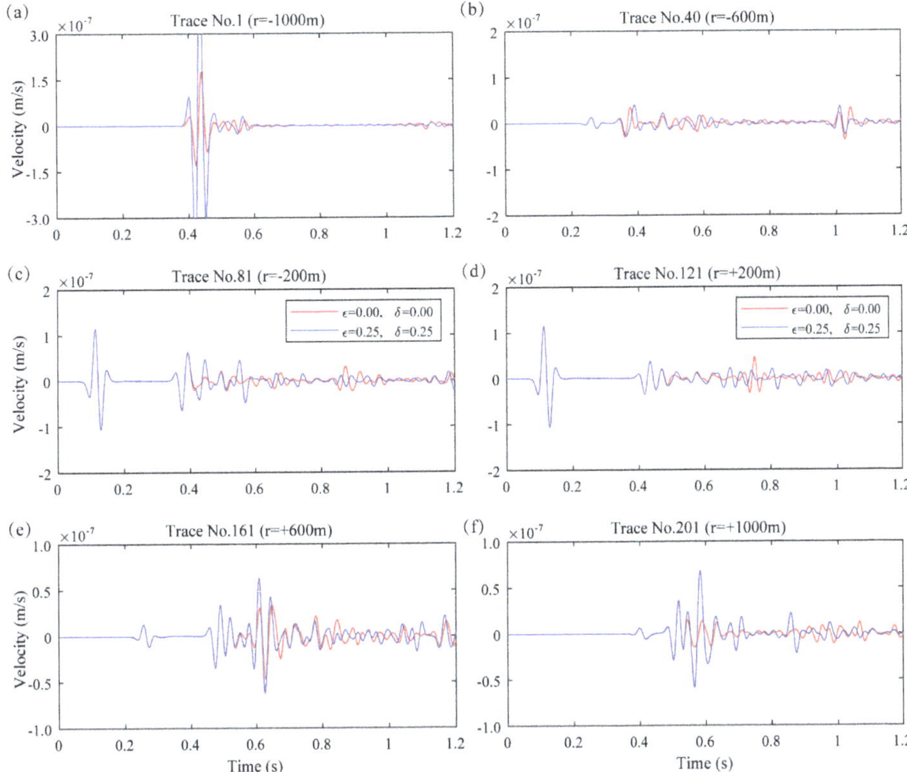

Figure 13. Comparison of Z-component waveforms from model M_{ED} and M_0 (with/without anisotropy, JY depression model). (**a**)–(**f**) are waveform records of geophones No.1/41/81/121/161/201 (with epicentral distance −1000/−600/−200/+200/+600/+1000 m, the sign indicates that the detectors are located on two sides of the source, positive along X-axis direction). Red indicates synthetic waveforms from model M_0 (without anisotropy); blue indicates seismic waveforms from model M_{ED} (with shale anisotropy).

Since the shape and structure of the JY model are very complex, it is difficult to analyze the impact of each seismic phase accurately. However, from Figures 12 and 13, we can clearly see that when anisotropy is added, in addition to the direct wave, the subsequent seismic phases are all affected and changed. Moreover, comparing the waveforms of detectors on the left and right side of the seismic source (located at trace No.101), it can be seen that their waveforms are not symmetrical and are affected by different degrees.

After the forward modeling of the JY Depression model, similar to Sections 3.1 and 3.2, we analyzed the waveforms' EM and PM of the anisotropic model M_E/M_D/M_{ED} from the model M_0 (without anisotropy). Figure 14 illustrates the calculation results. Here, the deviation results of Z- and X-components are still from the first survey line, while the Y-component results are from the second survey line (Figure 11). From Figure 14, we can find that:

- Similar to VTI and TTI model, the EM and PM of Z-component are still significantly larger than those of the X- and Y- horizontal components. The maximum EM is greater than one and the maximum PM reaches up to 0.5, which indicates that the impacts of anisotropy on the amplitude and phase are remarkable.

- Unlike the VTI and TTI model, for the JY model, the variation of EM and PM in the X- and Z-component with the epicentral distance (offset) is complicated, while the EM and PM of Y-component gets bigger with the increase of the epicentral distance.
- Same as TTI model but different from VTI model, for the JY model, the deviations of M_{ED} (ε = 0.25, δ = 0.25)/M_E (ε = 0.25, δ = 0)/M_D (ε = 0, δ = 0.25) are close to each other. This illustrates that the impact of different anisotropic parameters on the wavefield is complicated and needs further study.

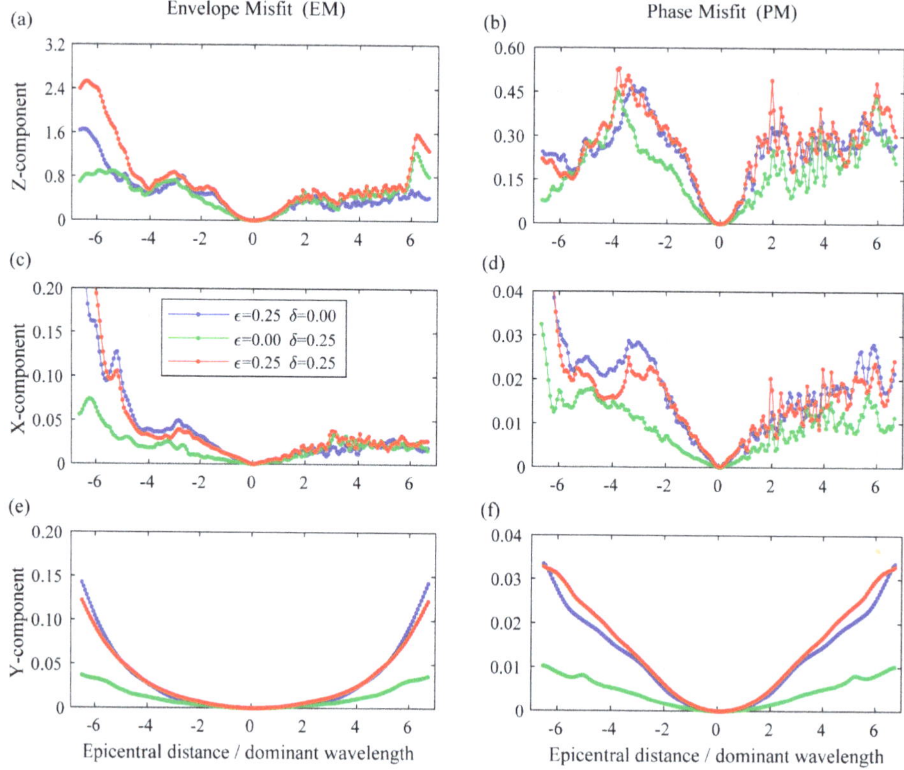

Figure 14. Envelope misfit and phase misfit of detectors' three-component waveforms (JY model). The meanings of each line and axis are similar to Figure 6.

5. Discussion and Conclusions

In order to study the impact of shale anisotropy on the seismic wavefield, we proposed a new quantitative evaluation method to calculate the waveform deviations of the anisotropic model from the regular model (without anisotropy). Based on the 3D elastic wave equation and the staggered-grid finite-difference method, the forward modeling theory of the three-component seismic wavefield considering the shale anisotropy was established. Then, we used the envelope misfit (EM) and the phase misfit (PM) to illustrate the differences in waveforms' amplitude and phase morphology caused by anisotropy. Finally, by comparing the waveforms of the models with/without anisotropy and calculating their EM and PM, we can quantitatively evaluate the impact of shale anisotropy on the seismic wavefield.

We applied the proposed method to analyze the anisotropy's effect on two simulation models (horizontal layered VTI model and curved layered TTI model) and one actual oilfield model (JY depression model). Tests on simulation and oilfield models prove that the proposed evaluation method is valid and efficient. Moreover, we gained some new cognitions about the anisotropy's effects. (1) The amplitude and phase deviations caused by anisotropy are significant, with EM lager than 1 and PM up to 0.5. Considering that the EM and PM are the relative deviations of two signals, this means that the shale anisotropy may cause the waveforms' amplitudes and phase shapes to be affected by a relatively high level. (2) Anisotropy's impacts on seismic waveform's vertical and horizontal components are different. In most cases, seismic surveys at the surface use the single-component (Z-component) detectors, so during the data interpretation, the Z-component records are adopted. We found that for these three different models, the anisotropy's effect on the Z-component is larger than the other X/Y components. (3) The impact of anisotropy on each detector depends on its offset from the source. When the geophones are set at the surface, for the horizontal layered VTI model, as the offset increases, the EM increases while the PM increases first then decreases. Moreover, the anisotropy's effects on two sides of the seismic source are symmetrical. However, for the curved layered TTI model and JY depression model, the variation trend is not clear. This illustrates that the anisotropy's impact relies heavily on the structure and symmetry axis of the medium model. (4) The influences of different anisotropy parameters on seismic wavefield are different. Based on the variable-controlling approach, we set three different anisotropic models (M_E/M_D/M_{ED}, whose ε/δ are 0.25/0, 0/0.25, and 0.25/0.25, respectively). For the VTI model, the deviations of M_{ED} from M_0 (without anisotropy) are the largest, while for the other two models (TTI and JY model), the deviations of M_E/M_D/M_{ED} are about at the same level.

According to our research, the impact of shale anisotropy on seismic wavefield is complicated and cannot be simply judged. So, for a particular oilfield model, we need specific analysis. The proposed method provides a useful and quantitative tool for anisotropy's evaluation. Furthermore, by using this method and calculating the waveform misfits from the actual seismic data, we can evaluate the accuracy of different anisotropic models and find the most suitable model for seismic data interpretation. Based on this idea, the anisotropic parameters inversion can be performed by seeking the minimum deviations between the synthetic and the actual seismic wavefield data. Moreover, as detectors with different offsets are affected differently, we can identify the observation area where the detectors are less affected by the shale anisotropy, then use the traditional interpretation method to process the selected wavefield data, so as to improve the accuracy and reliability of sweet spots prediction and reservoir characterization.

Author Contributions: All of the authors have contributed to the present paper. H.L. and X.C. conceived the method and directed the experiments. X.L. and J.L. gave advice for the numerical simulation and provided the model data of the actual oilfield. H.L. and R.W. performed the experiments. H.L. analyzed the data and wrote the paper.

Funding: This research was funded by the National Natural Science Foundation of China, grant number 41804050, 41974156, U1663207, and the State Energy Center for Shale Oil Research and Development, Sinopec, grant number G5800-18-ZS-KFNY002.

Conflicts of Interest: The authors declare no conflict of interest.

Appendix A. Staggered-Grid Finite-Difference Algorithm

In this paper, we used the staggered-grid finite-difference algorithm [21,22,26–28] to compute the seismic wavefield. This section gives a detailed introduction on this method. The 3D elastic wave equation can be written as:

$$\rho(x)\frac{\partial v_i}{\partial t}(x,t) = \sum_{j=1}^{3}\frac{\partial T_{ij}}{\partial x_j}(x,t) + \rho(x)F_i(x,t), \tag{A1}$$

where x and t are the space and time variables, subscript 1/2/3 refers to the X/Y/Z direction of space. $\rho(x)$ is the density and $v_i(x,t)$ is the particle velocity, $T_{ij}(x,t)$ is the stress tensor, and $F_i(x,t)$ is the external force. For the VTI (vertical transversely isotropic) medium:

$$\begin{bmatrix} \frac{\partial T_{11}}{\partial t} \\ \frac{\partial T_{22}}{\partial t} \\ \frac{\partial T_{33}}{\partial t} \\ \frac{\partial T_{23}}{\partial t} \\ \frac{\partial T_{13}}{\partial t} \\ \frac{\partial T_{12}}{\partial t} \end{bmatrix} = \begin{bmatrix} c_{11} & c_{12} & c_{13} & 0 & 0 & 0 \\ c_{12} & c_{11} & c_{13} & 0 & 0 & 0 \\ c_{13} & c_{13} & c_{33} & 0 & 0 & 0 \\ 0 & 0 & 0 & c_{44} & 0 & 0 \\ 0 & 0 & 0 & 0 & c_{44} & 0 \\ 0 & 0 & 0 & 0 & 0 & c_{66} \end{bmatrix} \begin{bmatrix} \frac{\partial v_1}{\partial x_1} \\ \frac{\partial v_2}{\partial x_2} \\ \frac{\partial v_3}{\partial x_3} \\ \frac{\partial v_2}{\partial x_3} + \frac{\partial v_3}{\partial x_2} \\ \frac{\partial v_1}{\partial x_3} + \frac{\partial v_3}{\partial x_1} \\ \frac{\partial v_1}{\partial x_2} + \frac{\partial v_2}{\partial x_1} \end{bmatrix}, \tag{A2}$$

in which c_{ijkl} is the stiffness matrix and can be represented by $v_{p_z}/v_{s_z}/\varepsilon/\delta/\gamma$ (see Equation (2) and Section 2.1). The staggered-grid finite-difference algorithm is built based on the following principle in order to approximate partial derivatives:

$$\frac{\partial f}{\partial a}(a) \approx \frac{1}{\Delta a}\sum_{m=1}^{N} k_m \times \left[f\left(a + \frac{2m-1}{2}\Delta a\right) - f\left(a - \frac{2m-1}{2}\Delta a\right)\right], \tag{A3}$$

where k_m is the finite difference coefficient (calculated by solving the Vandermonde matrix [33]) and N is the order number (we use the 5-th order during the forward modeling in this paper).

Following Equation (A3), the displacement velocity in Equation (A1) yields the following approximations (here shows v_1 for example):

$$\begin{aligned} &v_1(x_1,x_2,x_3,t+\Delta t) = v_1(x_1,x_2,x_3,t) + F_1\Delta t + \\ &\tfrac{\Delta t}{\Delta x}\tfrac{1}{\rho}\textstyle\sum_{m=1}^{N} k_m\Big[T_{11}\big(x_1 + \tfrac{2m-1}{2}\Delta x, x_2, x_3, t+\tfrac{1}{2}\Delta t\big) - T_{11}\big(x_1 - \tfrac{2m-1}{2}\Delta x, x_2, x_3, t+\tfrac{1}{2}\Delta t\big) + \\ &T_{12}\big(x_1, x_2 + \tfrac{2m-1}{2}\Delta x, x_3, t+\tfrac{1}{2}\Delta t\big) - T_{12}\big(x_1, x_2 - \tfrac{2m-1}{2}\Delta x, x_3, t+\tfrac{1}{2}\Delta t\big) + \\ &T_{13}\big(x_1, x_2, x_3 + \tfrac{2m-1}{2}\Delta x, t+\tfrac{1}{2}\Delta t\big) - T_{13}\big(x_1, x_2, x_3 - \tfrac{2m-1}{2}\Delta x, t+\tfrac{1}{2}\Delta t\big)\Big], \end{aligned} \tag{A4}$$

Following Equation (A3), the stress tensor in Equation (A2) yields the following approximations (here shows T_{11} for example):

$$\begin{aligned} &T_{11}(x_1,x_2,x_3,t+\Delta t) = T_{11}(x_1,x_2,x_3,t) + \\ &\tfrac{\Delta t}{\Delta x}\textstyle\sum_{m=1}^{N} k_m\Big[c_{11}\big[v_1\big(x_1 + \tfrac{2m-1}{2}\Delta x, x_2, x_3, t+\tfrac{1}{2}\Delta t\big) - v_{11}\big(x_1 - \tfrac{2m-1}{2}\Delta x, x_2, x_3, t+\tfrac{1}{2}\Delta t\big)\big] + \\ &c_{12}\big[v_2\big(x_1, x_2 + \tfrac{2m-1}{2}\Delta x, x_3, t+\tfrac{1}{2}\Delta t\big) - v_2\big(x_1, x_2 - \tfrac{2m-1}{2}\Delta x, x_3, t+\tfrac{1}{2}\Delta t\big)\big] + \\ &c_{13}\big[v_3\big(x_1, x_2, x_3 + \tfrac{2m-1}{2}\Delta x, t+\tfrac{1}{2}\Delta t\big) - v_3\big(x_1, x_2, x_3 - \tfrac{2m-1}{2}\Delta x, t+\tfrac{1}{2}\Delta t\big)\big]\Big], \end{aligned} \tag{A5}$$

in Equations (A4) and (A5), Δx and Δt are the spatial and time steps to approximate the spatial and time derivatives. For the staggered-grid finite-difference algorithm, the velocity and stress fields are staggered in time and space (Figure A1, [21,27,34]).

During the wavefield forward modeling, Δx and Δt should be chosen carefully in order to meet the stability condition:

$$\Delta t \le \frac{1}{\sqrt{3}} \cdot \frac{\Delta x}{v_{max}}, \quad \Delta x \le \frac{\lambda_{min}}{5} \tag{A6}$$

where v_{max} is the largest velocity in the medium and λ_{min} is the minimum wavelength of seismic wavefield. Moreover, when dealing with the boundary, we used the absorption boundary condition, i.e., applied the following attenuation function to the boundary area:

$$A(i) = \exp\left[-a * (i/N)^2\right], \tag{A7}$$

in which N and i are the total number and the serial number of grid points in the attenuation area, a is the attenuation coefficient (in this paper, $N = 100$, $a = 0.1$).

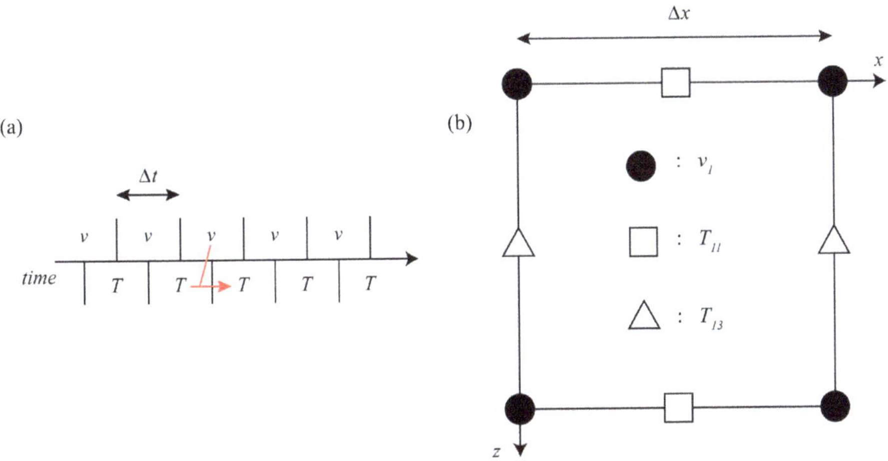

Figure A1. Schematic diagram of the staggered-grid finite-difference method. (**a**) Staggered grids in time; (**b**) staggered grids in space, v is the velocity and T is the stress, subscript 1 and 3 refer to the X and Z direction of space.

References

1. Zahid, S.; Bhatti, A.A.; Ahmad Khan, H.; Ahmad, T. Development of Unconventional Gas Resources: Stimulation Perspective. In Proceedings of the Production and Operations Symposium, Oklahoma City, OK, USA, 31 March–3 April 2007; Society of Petroleum Engineers: Dallas, TX, USA, 2007.
2. Jia, C.; Zheng, M.; Zhang, Y. Unconventional hydrocarbon resources in China and the prospect of exploration and development. *Pet. Explor. Dev.* **2012**, *39*, 139–146. [CrossRef]
3. Stark, P.H.; Chew, K.; Fryklund, B. The role of unconventional hydrocarbon resources in shaping the energy future. In Proceedings of the IPTC 2007 International Petroleum Technology Conference, Dubai, UAE, 4–6 December 2007; Volume 3, pp. 1876–1881.
4. Prasad, M.; Pal-Bathija, A.; Johnston, M.; Rydzy, M.; Batzle, M. Rock physics of the unconventional. *Lead. Edge* **2009**, *28*, 34–38. [CrossRef]
5. Vernik, L.; Nur, A. Ultrasonic velocity and anisotropy of hydrocarbon source rocks. *Geophysics* **1992**, *57*, 727–735. [CrossRef]
6. Johnston, J.E.; Christensen, N.I. Seismic anisotropy of shales. *J. Geophys. Res. Solid Earth* **1995**, *100*, 5991–6003. [CrossRef]
7. Grechka, V.; Yaskevich, S. Azimuthal anisotropy in microseismic monitoring: A Bakken case study. *Geophysics* **2014**, *79*, KS1–KS12. [CrossRef]
8. Ong, O.N.; Schmitt, D.R.; Kofman, R.S.; Haug, K. Static and dynamic pressure sensitivity anisotropy of a calcareous shale. *Geophys. Prospect.* **2016**, *64*, 875–897. [CrossRef]
9. Liu, X.-W.; Guo, Z.-Q.; Liu, C.; Liu, Y.-W. Anisotropy rock physics model for the Longmaxi shale gas reservoir, Sichuan Basin, China. *Appl. Geophys.* **2017**, *14*, 21–30. [CrossRef]
10. Wang, Y.; Li, C.H. Investigation of the P- and S-wave velocity anisotropy of a Longmaxi formation shale by real-time ultrasonic and mechanical experiments under uniaxial deformation. *J. Pet. Sci. Eng.* **2017**, *158*, 253–267. [CrossRef]
11. Zhang, F.; Li, X.; Qian, K. Estimation of anisotropy parameters for shale based on an improved rock physics model, part 1: Theory. *J. Geophys. Eng.* **2017**, *14*, 143–158. [CrossRef]
12. Zhang, Y.Y.; Jin, Z.J.; Chen, Y.Q.; Liu, X.W.; Han, L.; Jin, W.J. Pre-stack seismic density inversion in marine shale reservoirs in the southern Jiaoshiba area, Sichuan Basin, China. *Pet. Sci.* **2018**, *15*, 484–497. [CrossRef]

13. Carcione, J.M.; Helle, H.B.; Zhao, T. Effects of attenuation and anisotropy on reflection amplitude versus offset. *Geophysics* **1998**, *63*, 1652–1658. [CrossRef]
14. King, A.; Talebi, S. Anisotropy Effects on Microseismic Event Location. *Pure Appl. Geophys.* **2007**, *164*, 2141–2156. [CrossRef]
15. Warpinski, N.R.; Waltman, C.K.; Du, J.; Ma, Q. Anisotropy Effects in Microseismic Monitoring. In Proceedings of the SPE Annual Technical Conference and Exhibition, New Orleans, LA, USA, 4–7 October 2009; Society of Petroleum Engineers: Dallas, TX, USA, 2009.
16. Tsvankin, I.; Gaiser, J.; Grechka, V.; van der Baan, M.; Thomsen, L. Seismic anisotropy in exploration and reservoir characterization: An overview. *Geophysics* **2010**, *75*, 75A15–75A29. [CrossRef]
17. Sun, W.; Fu, L.; Guan, X.; Wei, W. A study on anisotropy of shale using seismic forward modeling in shale gas exploration. *Chin. J. Geophys.* **2013**, *56*, 961–970.
18. Meléndez-Martínez, J.; Schmitt, D.R. A comparative study of the anisotropic dynamic and static elastic moduli of unconventional reservoir shales: Implication for geomechanical investigations. *Geophysics* **2016**, *81*, D253–D269. [CrossRef]
19. Malehmir, R.; Schmitt, D.R. ARTc: Anisotropic reflectivity and transmissivity calculator. *Comput. Geosci.* **2016**, *93*, 114–126. [CrossRef]
20. Malehmir, R.; Schmitt, D.R. Acoustic Reflectivity From Variously Oriented Orthorhombic Media: Analogies to Seismic Responses From a Fractured Anisotropic Crust. *J. Geophys. Res. Solid Earth* **2017**, *122*, 10069–10085. [CrossRef]
21. Virieux, J. SH-wave propagation in heterogeneous media: Velocity-stress finite-difference method. *Geophysics* **1984**, *49*, 1933–1942. [CrossRef]
22. Virieux, J. P-SV wave propagation in heterogeneous media: Velocity-stress finite-difference method. *Geophysics* **1986**, *51*, 889–901. [CrossRef]
23. Qin, Y.; Zhang, Z.; Li, S. CDP mapping in tilted transversely isotropic (TTI) media. Part I: Method and effectiveness. *Geophys. Prospect.* **2003**, *51*, 315–324.
24. Guo, P.; McMechan, G.A. Sensitivity of 3D 3C synthetic seismograms to anisotropic attenuation and velocity in reservoir models. *Geophysics* **2017**, *82*, T79–T95. [CrossRef]
25. Thomsen, L. Weak elastic anisotropy. *Geophysics* **1986**, *51*, 1954–1966. [CrossRef]
26. Tsingas, C.; Vafidis, A.; Kanasewich, E.R. Elastic Wave Propagation in Transversely Isotropic Media Using Finite Differences. *Geophys. Prospect.* **1990**, *38*, 933–949. [CrossRef]
27. Graves, R.W. Simulating seismic wave propagation in 3D elastic media using staggered-grid finite differences. *Bull. Seismol. Soc. Am.* **1996**, *86*, 1091–1106.
28. Okaya, D.A.; McEvilly, T.V. Elastic wave propagation in anisotropic crustal material possessing arbitrary internal tilt. *Geophys. J. Int.* **2003**, *153*, 344–358. [CrossRef]
29. Wang, L.; Chang, X.; Wang, Y. Forward modeling of pseudo P waves in TTI medium using staggered grid. *Chin. J. Geophys.* **2016**, *59*, 1046–1058.
30. Kristek, J.; Moczo, P.; Archuleta, R.J. Efficient Methods to Simulate Planar Free Surface in the 3D 4th-Order Staggered-Grid Finite-Difference Schemes. *Stud. Geophys. Geod.* **2002**, *46*, 355–381. [CrossRef]
31. Danggo, M.Y.; Mungkasi, S. A staggered grid finite difference method for solving the elastic wave equations. *J. Phys. Conf. Ser.* **2017**, *909*, 0–5. [CrossRef]
32. Smith, J.O., III. *Mathematics of the Discrete Fourier Transform (DFT): With Audio Applications*, 2nd ed.; W3K Publishing: Stanford, CA, USA, 2007; ISBN 978-0974560748.
33. Liu, Y.; Sen, M.K. An implicit staggered-grid finite-difference method for seismic modelling. *Geophys. J. Int.* **2009**, *179*, 459–474. [CrossRef]
34. Bossy, E.; Talmant, M.; Laugier, P. Three-dimensional simulations of ultrasonic axial transmission velocity measurement on cortical bone models. *J. Acoust. Soc. Am.* **2004**, *115*, 2314–2324. [CrossRef]

© 2019 by the authors. Licensee MDPI, Basel, Switzerland. This article is an open access article distributed under the terms and conditions of the Creative Commons Attribution (CC BY) license (http://creativecommons.org/licenses/by/4.0/).

Article

High Throughput Screening and Characterization Methods of Jordanian Oil Shale as a Case Study

Ziad Abu El-Rub [1,*], Joanna Kujawa [2], Esra'a Albarahmieh [1], Nafisah Al-Rifai [1], Fathieh Qaimari [3] and Samer Al-Gharabli [1,*]

1. Pharmaceutical and Chemical Engineering Department, German Jordanian University, Amman 11180, Jordan
2. Faculty of Chemistry, Nicolaus Copernicus University in Toruń, 7 Gagarina Street, 87-100 Toruń, Poland
3. Chemical and Mineral Analysis Division, Jordan Ministry of Energy and Mineral Resources, P.O. Box 7, Amman 11118, Jordan

* Correspondence: ziad.abuelrub@gju.edu.jo (Z.A.E.-R.); samer.gharabli@gju.edu.jo (S.A.-G.); Tel.: +962-6-429-4412 (Z.A.E.-R.); +962-6-429-4404 (S.A.-G.)

Received: 7 July 2019; Accepted: 14 August 2019; Published: 16 August 2019

Abstract: Oil shale is an important possible solution to the problem of energy in Jordan. To explore the technical and the economic feasibility of oil shale deposits, numerous samples are analyzed using the standard Fischer Assay (FA) method. However, it would be useful to develop faster, cheaper, and reliable methods for determining the oil content of oil shale. Therefore, the aim of this work was to propose and investigate rapid analytical techniques for the screening of oil shale deposits and to correlate them with the FA method. The Omari deposit located east of Jordan was selected as a case study for analysis using thermogravimetric analysis (TGA) coupled with Fourier-transform infrared (FTIR), differential scanning calorimetry (DSC), elemental analysis, X-ray fluorescence (XRF), X-ray diffraction (XRD), scanning electron microscopy (SEM), and energy-dispersive X-ray (EDX) analysis. Results obtained from the TGA method were linearly correlated with FA with high regression factor (R^2 = 0.99); a quadratic correlation (R^2 = 0.98) was maintained between the FA and the elemental hydrogen mass content, and a quadratic correlation (R^2 = 0.97) was found between the FA and the aliphatic hydrocarbons (FTIR peak at 2927 cm^{-1}) produced in the pyrolysis zone. Although other techniques were less correlated, further investigation might lead to better results. Subsequently, these correlated techniques can be a practical alternative to the conventional FA method when, in particular, specific correlation is made for each deposit.

Keywords: oil shale; Jordan; TGA; FTIR; DSC; elemental analysis; XRF; XRD; SEM; EDX

1. Introduction

Jordan is a country in the Middle East region that suffers from a lack of conventional fossil energy sources. In 2017, the country imported 94% of its energy needs, equivalent to 8.5% of its gross domestic product (GDP) [1]. To increase energy security, the kingdom has been encouraging investments in renewable energy—mainly solar and wind—and alternative energy that is principally oil shale. However, the economic feasibility of these projects is affected by crude oil prices, technical challenges, and environmental considerations.

Oil shale can be defined as sedimentary rocks including organic matter disseminated in a mineral matrix. The organic matter is combustible hydrocarbons, predominantly kerogen, which is insoluble in common organic solvents [2,3].

The kingdom is ranked the sixth country in the world in oil shale resources, with more than 70 billion tons of proven reserves distributed over 60% of Jordan's territory [4]. Recent scientific reports highlighted that Jordan can produce around 34 billion barrels of shale oil from its resources [5]. Despite the fact that these significant amounts were discovered in the early 20th century, Jordan's experience in exploiting oil shale commercially is relatively modest. The first commercial oil shale project in Jordan will be the exploitation of Attarat Umm Ghudran deposit by direct burning to produce 470 MW electricity by the year 2020. Furthermore, subsurface in-situ retorting has been investigated by Jordan Oil Shale Company (JOSCO), which is a wholly-owned subsidiary of Royal Dutch Shell.

Jordan has given international companies the license of commercial exploitation for several oil shale deposits. These deposits are distributed at more than 24 locations across Jordan, as shown in Figure 1. However, most commercial areas are in the central region of Jordan (El-Lajjun, Sultani, Attarat Um Ghudran, and Wadi Maghar) [4,6]. The possibility for commercially exploiting a deposit depends on the applied technology and the type of application. Generally, the oil shale can conventionally be exploited by either retorting to produce oil or by direct combustion to generate electricity. From an economic perspective, the retorting process requires an oil content of more than 5 wt.%, whereas the direct burning process requires an oil shale minimum calorific value of 3 MJ/kg [7]. The Jordanian oil shale is suitable for both applications, as it has an average oil content of 10 wt.% and a minimum calorific value of 6 MJ/kg. Moreover, the average sulfur content in the Jordanian oil shale is about 3 wt.% of the rock weight or 9 wt.% relative to the kerogen content. The presence of sulfur and other compounds such as nitrogen, oxygen-containing compounds, and diolefins reduces the stability and the quality of the produced shale oil, which complicates the shale oil treatment process and decreases economic feasibility [7]. Therefore, this shale oil needs to be upgraded to produce stable liquid fuel [2]. Upgraded oil shale can be utilized more efficiently and economically in power generation as well as in oil production [8].

Oil shale deposits usually have large areas, different depths, and varying oil content. Therefore, to exploit such complex resources, it becomes critical to find high throughput screening methods. There are a variety of physical and chemical methods that are used to screen and characterize oil shale [9,10]. These methods include solvent extraction, microscopic methods [11], physical (color, specific gravity, thermophysical properties), elemental analysis (C, H, N, O, S), thermal degradation (Fischer Assay, thermogravimetric analysis, isothermal pyrolysis, pyrolysis flame ionization detection) [12,13], chemical degradation, chromatographic procedures, mass spectrometry, Raman spectroscopy [14], laser-induced techniques [15], and modern infrared techniques [16]. However, each method has its own shortcomings that limit its application. Thus, the choice of the optimal method depends on cost, accuracy, reliability, speed, online capability, and level of operator intervention.

El-Hasan et al. [5] presented the utilization of synchrotron-supported techniques to analyze Jordanian oil shale deposits. Synchrotron-based X-ray absorption near-edge structure (XANES) was used for chromium and arsenic analysis [5]. For the speciation analysis of solids focused on trace elements in environmental samples, synchrotron beamlines have the potential of competing in X-ray diffraction, fluorescence, as well as absorption spectroscopy [17]. Considering the study of kerogens structure, cross-polarized magic angle spinning (CP/MAS) and X-ray photoelectron spectroscopy (XPS) techniques are very useful. ^{13}C NMR was utilized with high efficiency to determine the carbon skeletal structure in fossil resources [18,19]. XPS, however, was successfully implemented in the investigation of the forms of organic heteroatoms (O, S, N) in the complex matrix [20,21]. The Fischer Assay (FA) method is the standard test method for determining the oil yield in the oil shale deposit [22–25]. Despite being the standard method, it has several disadvantages, including inability to measure the maximum amount of oil in a given oil shale [26], expensive laboratory setup, strict laboratory safety measures due to evolved H_2S and other toxic gases, relatively large sample size (100 g), prolonged analysis time (~2 h), dependency on retort parameters (particle size, heating rate, heat and mass transfer limitations, etc.), and lower yield compared with commercial retorts [10]. Therefore, it would be advantageous to develop faster, cheaper, and more reliable methods for determining the oil content of oil shale.

The aim of this study was to find high throughput screening methods for analyzing the oil content of oil shale as an alternative for the conventional FA method. The Omari oil shale deposit in Jordan was characterized and adopted as a case study. A variety of techniques were screened for this purpose, including thermogravimetric analysis (TGA) coupled with Fourier-transform infrared (FTIR), differential scanning calorimetry (DSC), X-ray diffraction (XRD), scanning electron microscopy (SEM), and energy-dispersive X-ray (EDX).

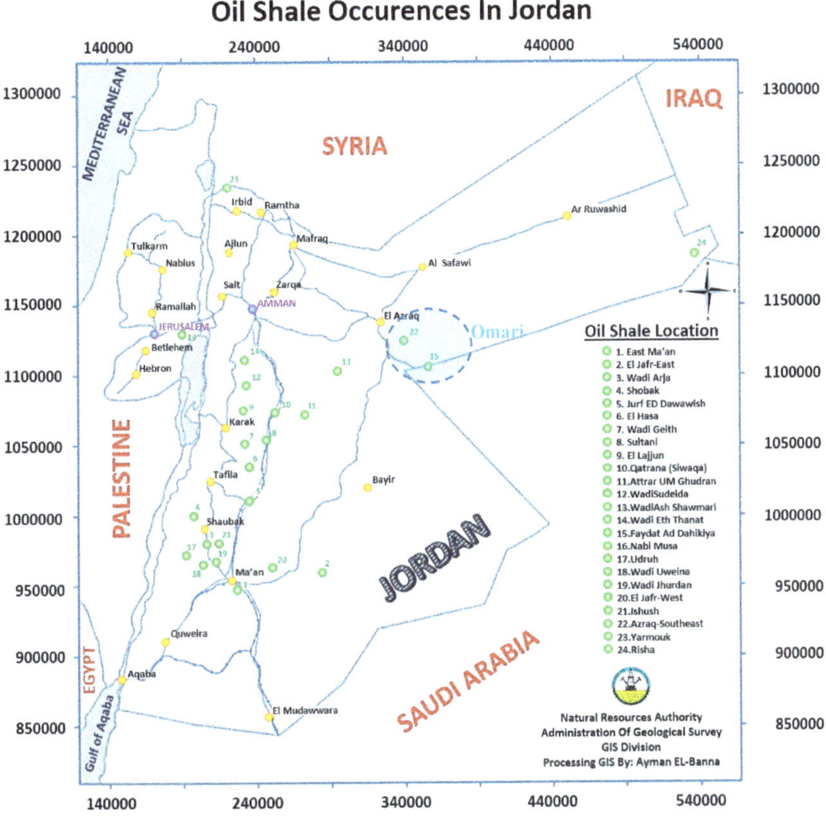

Figure 1. Map of the major oil shale deposits in Jordan. Adapted from reference [4].

2. Materials and Methods

Crushed and sieved oil shale samples were collected from the Omari deposit, which is located southeast of Amman and nearly reaches the Iraqi borders, as shown in Figure 1. Moreover, the oil yield was determined using the FA method according to standard test method ASTM D3094-90. In this method, 100 g of a dried sample of oil shale was ground into less than 2.38 mm particles and heated in a small aluminum oven (retort) to 500 °C at a rate of 12 °C/min and held at the final temperature for 40 min. The resulting vapors of oil, gas, and water were passed through a condenser and cooled with ice water into a graduated centrifuge tube to separate the oil from the water and measure the oil yield [27].

Thermal properties of the samples were evaluated using thermoanalyzer Jupiter STA 449 F5 (Netzsch, Germany) coupled with FTIR spectrometer (Vertex 70V Bruker Optik, Germany). The different samples obtained from the deposit were placed in vials and given a number from OS1 to OS6. Three samples from each vial were thermally analyzed. We assumed that these homogeneous and representative samples could be considered as an identical sample where the standard deviation of the thermal analysis is calculated. Thermogravimetric analysis (TGA), derivative thermogravimetry (DTG), and differential thermal analysis (DTA) were accomplished over a temperature range of 25–1100 °C under an ambient atmosphere of nitrogen with a heating rate of 20 °C/min. TGA and DTG were used for the thermal stability characterization of the investigated samples. DTA gives information about the changes in the energy of the materials (specific heat capacity and enthalpy). The gas products from TGA were simultaneously analyzed on the FTIR spectrometer with a resolution of 4 cm^{-1} and a number of 16 scans/0.3 °C by the implementation of Gram-Shmidt diagram.

Elemental analysis was done on Vario MACRO CHN ELEMENTAR Analysensysteme GmbH. Furthermore, the chemical composition of major oxides in the oil shale samples was analyzed by XRF using the sequential wavelength dispersive X-ray fluorescence spectrometer (Bruker S4 Pioneer, Germany). The identification of the structure of the crystalline solid shales and their mineral composition was examined by using the D4 ENDEAVOR X-ray Diffractometer from Bruker. The method of peak area against calibrated curves was used.

The Quantax 200 with XFlash 4010 detector (Bruker AXS machine, Czech Republic) was used for scanning electron microscopy (SEM) imaging of samples. SEM imaging was performed with a secondary electrons (SE) detector under high vacuum $<6 \times 10^{-4}$ Pa. Before measurements, oil shale samples were sputtered with a nanolayer (10 nm) of gold for morphology determination. For EDX measurements, the microscope 1430 VP LEO Electron Microscopy Ltd (England) was used (non-sputter samples were tested).

3. Results and Discussion

The absence of an ideal method to analyze oil shale has led to the usage of a variety of methods to overcome various limitations of the popular method(s) of analysis. The results of these analyses are discussed here.

3.1. Fischer Assay

Six oil shale samples labeled from OS1 to OS6 were obtained from the Omari deposit covering a wide range of oil content in Jordanian oil shale, which has an average content of about 10 wt.% [7]. The oil yield was determined using the standard FA method. Table 1 shows that the oil content of the tested samples varied between 2.9 to 12.7 wt.%, and the oil density varied between 0.951 to 0.967 g/mL. Furthermore, it was noticeable that the shale oil content was inversely proportional to its density, which could be related to the lower density of oil compared with the inorganic matter.

Table 1. Fischer Assay (FA) analysis of oil shale (OS) samples.

Sample	Fischer Assay		OS Density
	(g/100 g)	(mL/100 g)	(g/mL)
OS1	2.90	3.00	0.967
OS2	5.31	5.50	0.964
OS3	7.00	7.25	0.966
OS4	9.10	9.48	0.960
OS5	11.20	11.70	0.957
OS6	12.70	13.35	0.951

3.2. Thermogravimetric Analysis

TGA is a rapid technique widely used to record weight loss of oil shale as a function of temperature. DTA is a complementary technique usually used in the same TGA instrument to represent the temperature difference developed between the oil shale sample and a reference. Both techniques can be used to provide important information on the behavior of heated oil shale samples and assign the maximum temperature where the basic thermodecomposition of kerogen occurs. Moreover, thermogravimetric analysis coupled with FTIR (TGA-FTIR) as a detector was used for online identification of evolved gases of the TGA in the infrared spectrometer gas cell. This coupling can be useful in studying the mechanism of kerogen pyrolysis, which gives a real-time identification of TGA products at different temperatures. Furthermore, gases evolved at the temperature observed for oil weight loss (TGA/DTG) were coupled with the thermal transitions gained by coupling to the DSC. The simultaneous DSC analysis was also associated with endothermic transitions around the region of gas evolving from oil pyrolysis. Additionally, the variation of the amount of evolved gases during pyrolysis and thermal degradation of the oil shale sample could be detected by Gram-Schmidt thermogram where the FTIR absorption intensity of evolved gases was plotted as a function of temperature.

3.2.1. TGA and DTG

All six oil shale samples collected from Omari deposit were subjected to the above-mentioned techniques under a nitrogen environment over a temperature range of 25 to 1100 °C and at a heating rate of 20 °C/min. In general, all samples showed similar trends with slight discrepancies in the pyrolysis region. Figure 2A shows representative TGA and DTG curves of one of the oil shales sample (OS6, FA 12.70 wt.%). The total weight loss was 52.0 wt.%, which passed through three distinct stages: (1) drying, (2) pyrolysis, and (3) mineral decomposition. In the drying stage (up to 370 °C), less than 1.5 wt.% was lost due to moisture loss, which included surface moisture, shirking core evaporation of moisture, and the moisture in small capillaries of the oil shale particles [2]. In the pyrolysis stage (370–560 °C), a weight loss of 14.0 wt.% was recorded. Here, kerogen was thermally decomposed to form shale oil, shale gas, and char. Furthermore, the mineral matter may have decomposed to produce carbon dioxide and release combined water. Finally, in the mineral decomposition stage (more than 600 °C), about 31.0 wt.% loss was observed. In this stage, the different minerals, e.g., calcite, kaolinite, and dolomite, decomposed to form mainly carbon dioxide and the metal oxide. Moreover, employing the first derivative for the TGA to generate the corresponding DTG showed an average maximum temperature of pyrolysis (T_{max}) of 472.4 ± 9.1 °C.

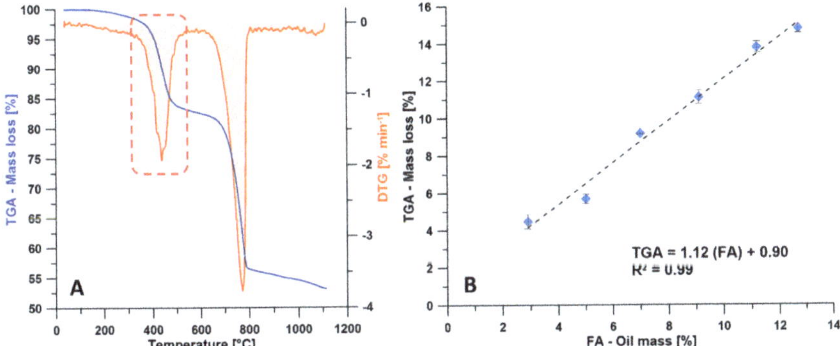

Figure 2. (**A**) Representative decomposition curve of OS6 (FA 12.7 wt.%) using thermogravimetric analysis (under nitrogen). The measured values of decomposition and oil release around 435 °C were calculated and optimized through comparison of the derivative thermogravimetry (DTG) curves at a heating rate of 20 °C/min; the solid line represents the thermogravimetric analysis (TGA) curve, and the dashed line represents the first derivative thermogravimetric (TG) curve. (**B**) Correlation curve with corresponding oil content in wt.% from Fischer Assay plotted against the measured oil using thermogravimetric analysis showed a near-perfect linear fitting for oil content detection from 2.9 wt.% to 12.7 wt.%.

When correlating the TGA mass loss in the pyrolysis region (370–560 °C) with the FA oil content, an outstanding linear correlation (R^2 = 0.99) between the two analysis methods was established (Figure 2B). Each point in Figure 2B represents an average of three trials, and the corresponding error bars demonstrate the precision of the experiments. Therefore, the TGA method can be used as a reliable alternative for the FA method. Because kerogen type and mineral matrix might vary according to the deposit type [9], such correlation should be utilized based on a specific calibration. Moreover, the produced correlation in Figure 2B can be also described as in Equation (1).

$$FA = 0.894 \times TGA - 0.802 \tag{1}$$

where FA denotes the oil weight percent in the oil shale determined by the FA method, and TGA represents the percent of weight loss at the pyrolysis range. Such correlation is in accordance with the literature [28].

3.2.2. TGA-FTIR

Thermogravimetric analysis coupled with FTIR as a detector was used for online identification of evolved gases of TGA for all samples (OS1 to OS6). The analysis was repeated three times in the infrared spectrometer gas cell, and the average result for each sample is presented in Figure 3. The spectra showed two distinct regions: pyrolysis and mineral decomposition. During the pyrolysis region (370–560 °C), the online FTIR spectrum showed broad bands between 2793–3060 cm^{-1} assigned to aliphatic (C-H stretching) with a peak area that was directly proportional to oil content. It is noteworthy to mention that CO_2 also evolved at this stage (pyrolysis), which was confirmed by the broad bands between 2225–2410 cm^{-1} assigned to CO_2 stretching with no obvious correlation related to the oil content of the samples. Furthermore, the dactyloscopic region (500–1500 cm^{-1}) showed the characteristic peaks related to the inorganic and the organic matrix. In the second region (thermal decomposition), weak bands with respect to C-H aliphatic stretching (2793–3060 cm^{-1}) were shown, which could have been related to the low oil content remaining after pyrolysis. On the other hand, stronger bands were noticed for CO_2 (2225–2410 cm^{-1}) due to the thermal decomposition of minerals, which produced CO_2.

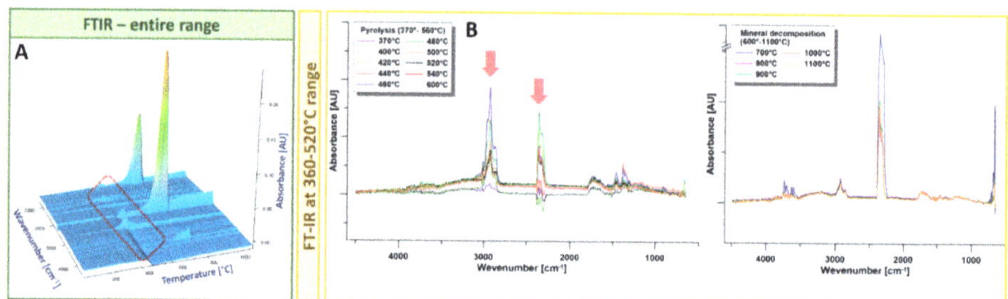

Figure 3. (**A**) 3D FTIR spectra collected as the function of temperature. (**B**) Detailed analysis of FTIR spectra of pyrolysis and mineral decomposition regions for a selected oil shale sample (OS6).

Figure 4A shows the 3D FTIR spectra between wavenumbers (2800–300 cm^{-1}) for the six oil shale samples at the pyrolysis temperature range (370–560 °C). Quantification was done based on the integration of peak areas at 2927 cm^{-1} and 2342 cm^{-1} for the aliphatic hydrocarbons and CO_2, respectively. Moreover, a correlation was made between the FA and the areas of the two sets of peaks, as shown in Figure 4B. It was obvious that a good quadratic correlation (R^2 = 0.97) was maintained for the aliphatic hydrocarbons, as shown in Equation (2).

$$Y = 0.01009 \times X^2 + 0.02401 \times X + 0.10568 \tag{2}$$

where Y is the area of aliphatic hydrocarbons (peak at 2927 cm^{-1}), and X is the oil content of FA (wt.%).

On the other hand, there was no clear correlation between the FA and the peak area of CO_2 (2342 cm^{-1}). However, two regions representing two mechanisms could be envisaged where there was a certain production of CO_2 up to FA 7.0 wt.% and then it dropped.

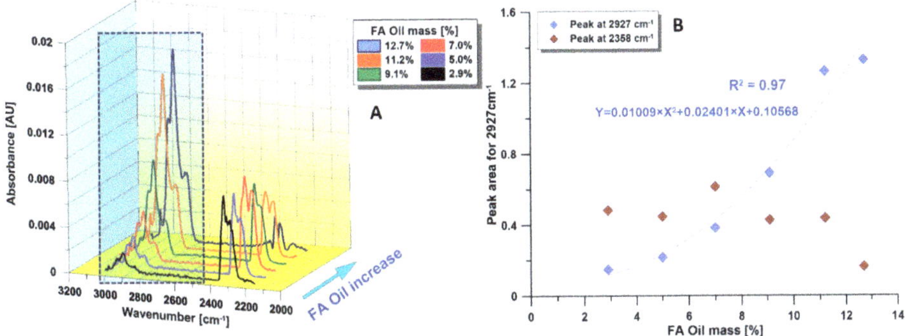

Figure 4. (**A**) Fourier-transform infrared (FTIR) spectra of the six oil shale samples at a temperature range of 370–560 °C and a heating rate of 20 °C/min, thus representing the major evolved gases. (**B**) Correlation curve between oil content in wt.% from FA and the measured oil using dynamic FTIR at 2927 cm^{-1} and the measured CO_2 at 2342 cm^{-1}.

3.2.3. DSC

The amount of heat needed for retorting oil shale depends predominantly on its specific heat capacity. Therefore, DSC analysis was performed for all samples in triplicate. Figure 5 shows, using OS6 (FA 12.7%) as an example, the DSC curve in addition to its TGA and DTG profiles. Results showed a similar DSC profile as a function of temperature for all samples. Furthermore, the peaks denoted an endothermic behavior in the pyrolysis zone (370–560 °C) with a heat capacity in the range of (0.133–1.228 J/(g·K)) due to the thermal decomposition of kerogen. Moreover, the specific heat of oil shale depended on its type, which differed in the composition of mineral and kerogen content. However, for the same type, the specific heat increased with higher oil yield [2]. The same trend was found for all oil shale samples (OS1 to OS6), as exemplified in Figure 6, where specific heat capacity (Figure 6A) and enthalpy (Figure 6B) were directly proportional to the oil content.

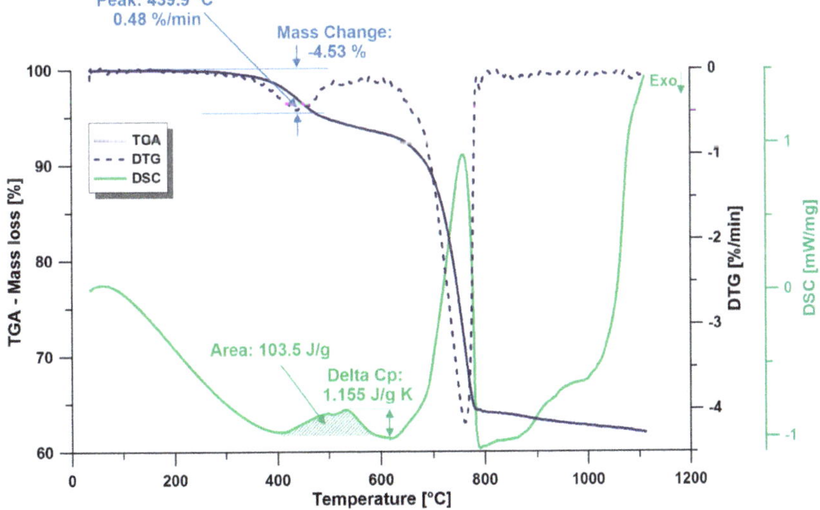

Figure 5. Representative TGA (mass loss) and DTG (mass change rate) curves for OS6 at a heating rate of 20 °C/min (solid blue and dashed blue lines, respectively) overlapped with differential scanning calorimetry (DSC) curve (solid green line) at a heating rate of 20 °C/min. Left-hand side *y*-axis: TG curve. Right-hand side axis: DTG and DSC curves.

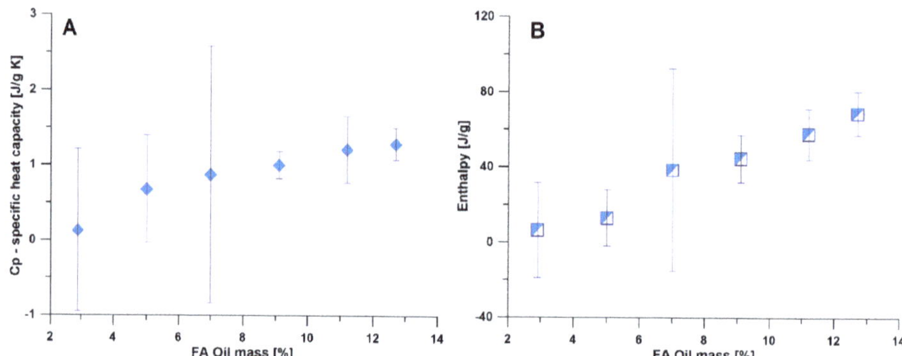

Figure 6. The relationship between the FA oil yield and (**A**) specific heat capacity and (**B**) enthalpy.

3.2.4. Gram-Schmidt

The amount of evolved gases resulted from the TGA could be detected and evaluated using the Gram-Schmidt thermogram [29]. For OS1 to OS6, Gram-Schmidt showed increase in the amount of evolved gases as a function of oil content, where the maximum gas productions were around 450 °C and 760 °C for the kerogen pyrolysis and the mineral decomposition, respectively (Figure 7, OS6 at a heating rate of 20 °C/min as an example). This behavior mirrored the peaks found in the DTG shown in Figure 2A. Figure 8 shows two sets of oil shales. The first group represents the low oil content (OS1–OS3), and the second is the high oil content (OS4–OS6). Although the first group contained lower oil content, it showed a higher amount of evolved gases. This might have been related to their lower heat capacity required by the oil for pyrolysis (as shown in Figure 6). Conversely, the second group produced fewer gases due to their higher heat capacity.

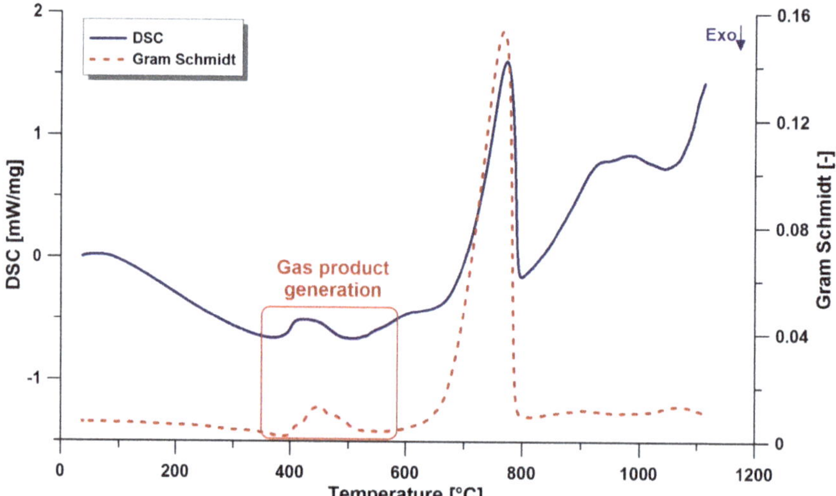

Figure 7. Triplicate heat flow rate (DSC) curve (solid line) and Gram-Schmidt (GS) signal (dashed line) of a representative OS6 (12.7 wt.%), showing the region of gas evolving in the temperature range of 370–560 °C determined as the peak temperature. Heating rate of 20 °C/min, gas atmosphere: nitrogen.

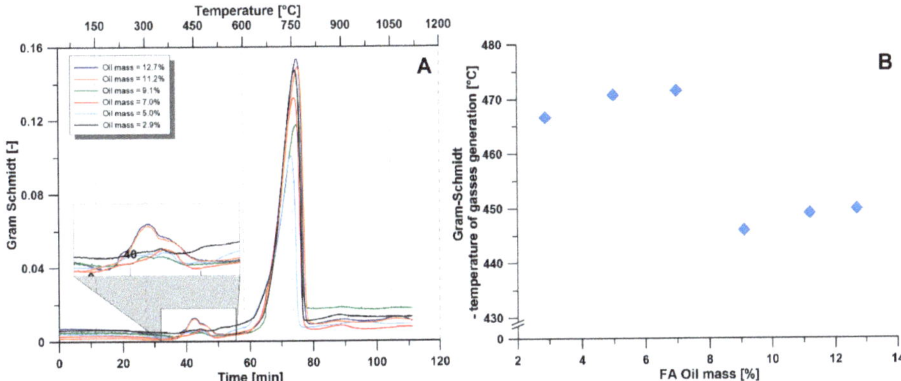

Figure 8. (**A**) Gram-Schmidt absorption intensity of evolved gases for the samples containing different levels of oil content. (**B**) Correlation of FA and Gram-Schmidt (temperature of gases generation).

3.3. Elemental and Mineral Analysis

Elemental analysis is an accurate standard method used for determining the carbon-hydrogen-nitrogen (CHN) content of kerogens by combustion. However, it could be used for determining the origin and the evolution of oil shale with limited accuracy [2]. XRF was conducted to determine the major oxides present in the oil shale samples, whereas XRD was used as a qualitative analytical method for the identification of the crystal phase of minerals in the oil shale samples. SEM revealed information about the solid shale including outer morphology, chemical composition, crystalline structure, and orientation of materials making up the sample. EDX was one of the analytical techniques used for the elemental analysis or the chemical characterization of a sample. The interaction of a source of X-ray excitation and a sample gave a unique atomic structure for each element, which allowed a unique set of peaks on the electromagnetic emission spectrum.

3.3.1. Elemental analysis

CHN analysis was conducted for all oil shale samples (OS1 to OS6). Figure 9A shows a correlation between the FA (wt.%) and the mass content of each element, where a directly proportional relationship between the two was observed. Out of the three possible correlations, elemental hydrogen mass content provided the best correlation to the FA with a good quadratic correlation (R^2 = 0.98), as shown in Equation (3). Correlating the hydrogen content to the FA method is preferred because it is predominantly found in the organic matter (kerogen), while carbon is found in both inorganic and organic fractions.

$$H = 0.0208 \times FA^2 - 0.1849 \times FA + 1.2392 \quad (3)$$

where H represents the weight percent of hydrogen content in the oil shale sample, and FA denotes the oil weight percent in the oil shale determined by the FA method.

Furthermore, the atomic ratios H/C and N/C were calculated and evaluated against the oil content. There was a relatively fixed N/C atomic ratio (0.01–0.02), whereas a wider range of H/C atomic ratio (0.69–1.12) was found. Values of the H/C ratio could be used as a preliminary indicator of the aromatic nature of kerogen. However, the atomic ratios of H/C and O/C of the kerogen content (not the oil shale) could be assigned on the van Krevelen diagram to determine the type of kerogen [10]. It is reported that the kerogen of Jordanian oil shale has an average content of 10 wt.% and atomic H/C and O/C ratios of 1.3 and 0.101, respectively [2].

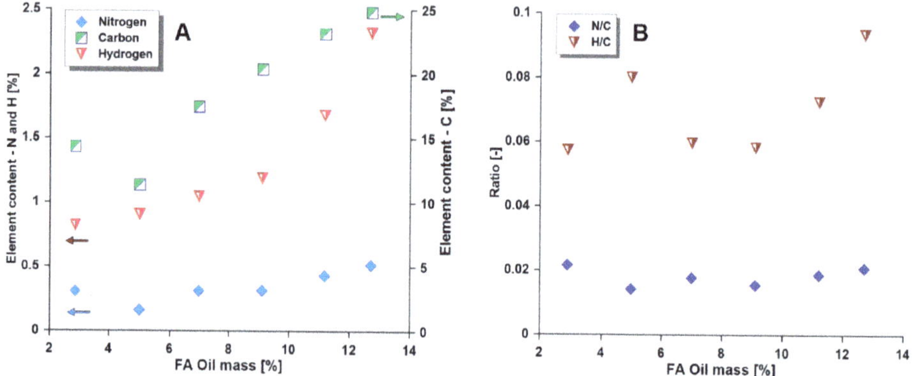

Figure 9. Relation between oil content in wt.% from Fischer Assay and the (**A**) elemental analysis (wt.%) of C, H, and N of solid oil shale samples, (**B**) atomic ratios of H/C and N/C of the solid oil shale samples.

3.3.2. XRF

The XRF analysis of the mineral matter resulted from tested oil shale samples, as shown in Table 2, indicated that this oil shale contained mainly calcite ($CaCO_3$) and quartz (SiO_2). Furthermore, the data showed no trend in the mineral composition that could be correlated with the oil content. The mineral content of the tested samples agreed with the reported average mass concentrations of major inorganic content of oil shale deposits in the central part of Jordan [7].

Table 2. Comparative X-ray fluorescence (XRF) chemical composition in wt.% of investigated oil shale samples with different oil contents.

Sample	Shale Oil (wt.%)	CaO	SiO_2	Al_2O_3	P_2O_5	Fe_2O_3	MgO	K_2O	TiO_2
OS1	2.9	41.0	10.7	3.83	1.77	1.71	0.58	0.34	0.11
OS2	5.0	21.4	41.3	2.35	3.81	1.30	2.83	0.77	0.12
OS3	7.0	44.6	3.97	1.19	8.51	0.61	0.12	0.01	0.02
OS4	9.1	37.2	11.2	2.05	1.13	1.09	0.33	0.08	0.08
OS5	11.2	36.9	7.76	1.21	0.83	0.73	0.39	0.09	0.05
OS6	12.7	33.8	9.28	2.46	1.21	1.19	0.63	0.28	0.11

3.3.3. XRD

XRD analysis of all oil shale samples (OS1 to OS6) showed the same type of minerals with different concentrations. In Figure 10, OS6 (FA 12.7 wt.%) was used as an example to reveal the presence of the main minerals in the oil shale samples, which were quartz (SiO_2), ankerite ($Ca(Fe,Mg)(CO_3)_2$), calcite, apatite ($Ca_{10}(PO_4)_6F_2$), and clay ($Mg\text{-}SiO_2\text{-}OH\text{-}H_2O$).

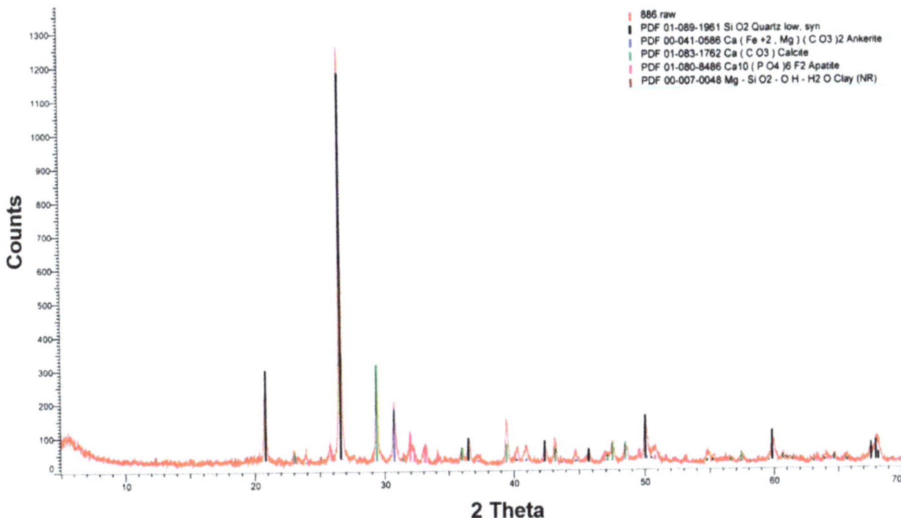

Figure 10. XRD analysis of raw oil shale sample (OS6 FA 12.7 wt.%).

3.3.4. SEM

Figure 11 shows scanning electron micrographs for selected oil shale samples that covered the whole range of oil content given the symbols A (OS1 2.9 wt.%), B (OS3 7.0 wt.%), and C (OS6 12.7 wt.%) with variable magnification from 1 kx (magnification—1000 times) to 35 kx (magnification—35,000 times). The three samples showed amorphous and fragmented surfaces with no distinct clusters of organic materials. EDX analysis (Figure 12) showed the same chemical components detected by XRF with no obvious trend that could be correlated with oil content, which might have been related to the anisotropy of the surface. Figure 13 shows the micrograph with colored mineral distribution mapped on the surface of oil shale and quantified at a rate of 10.67 Kcps in a pulsed mode and multi-line scanning. It could be observed that the following elements, C, O, Mg, Al, Si, S, K, Ca, Fe, Cl, and Zn, were uniformly distributed across the whole surface. Conversely, phosphorus (P) was located in certain spots with less distribution because it originated from the apatite, as found in the XRD analysis. Calcium then silicon and sulfur were found at the highest level, which is in accordance with the literature values of Jordanian oil shale deposits (30–35% $CaCO_3$, 10–26.3% SiO_2, 4.3–4.8% SO_3) [7]. On the other hand, a more precise line scan analysis was conducted and is presented in Figure 14. The line analysis presented the same mineral types found by XRD and XRF analyses and showed no localization of the organic matter. Moreover, it demonstrated the diversity of mineral content and that there was no apparent relationship with the oil content.

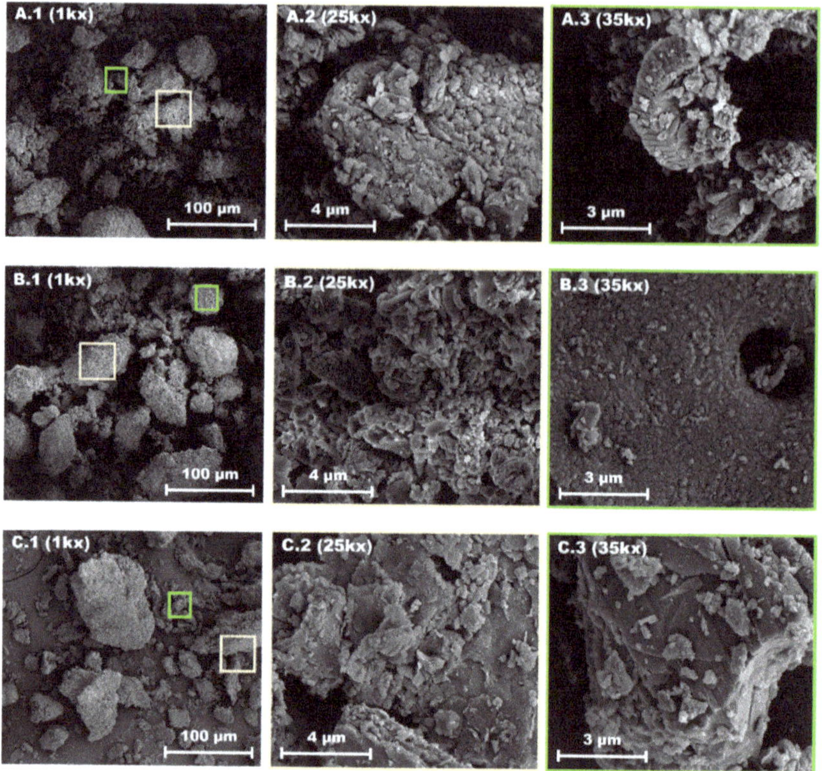

Figure 11. Scanning electron microscope (SEM) photos of samples containing **A**—2.9 wt.%, **B**—7.0 wt.%, and **C**—12.7 wt.% of oil shale.

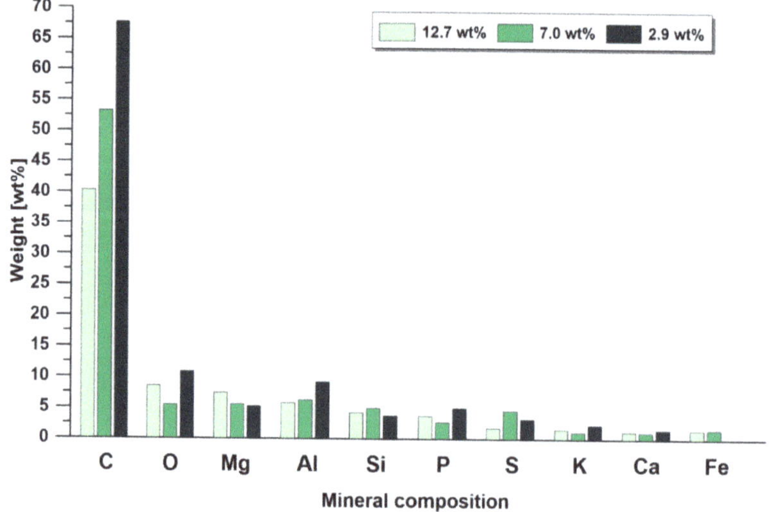

Figure 12. Comparative mass chemical composition of investigated oil shale samples with different oil content as determined by SEM.

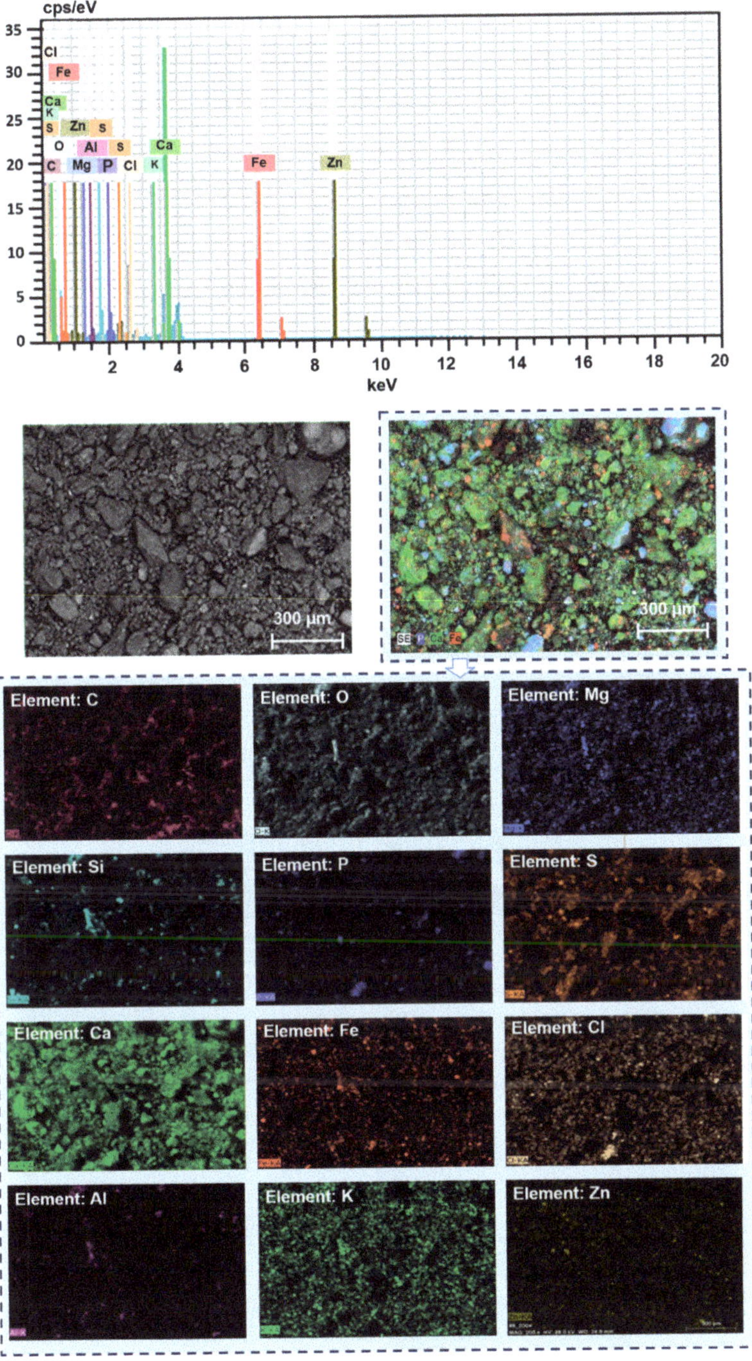

Figure 13. SEM identification of the chemical composition of OS3 with medium oil content (FA 7 wt.%), showing micrograph with colored mineral distribution mapped and quantified at a rate of 10.67 Kcps in a pulsed mode and multi-line scanning.

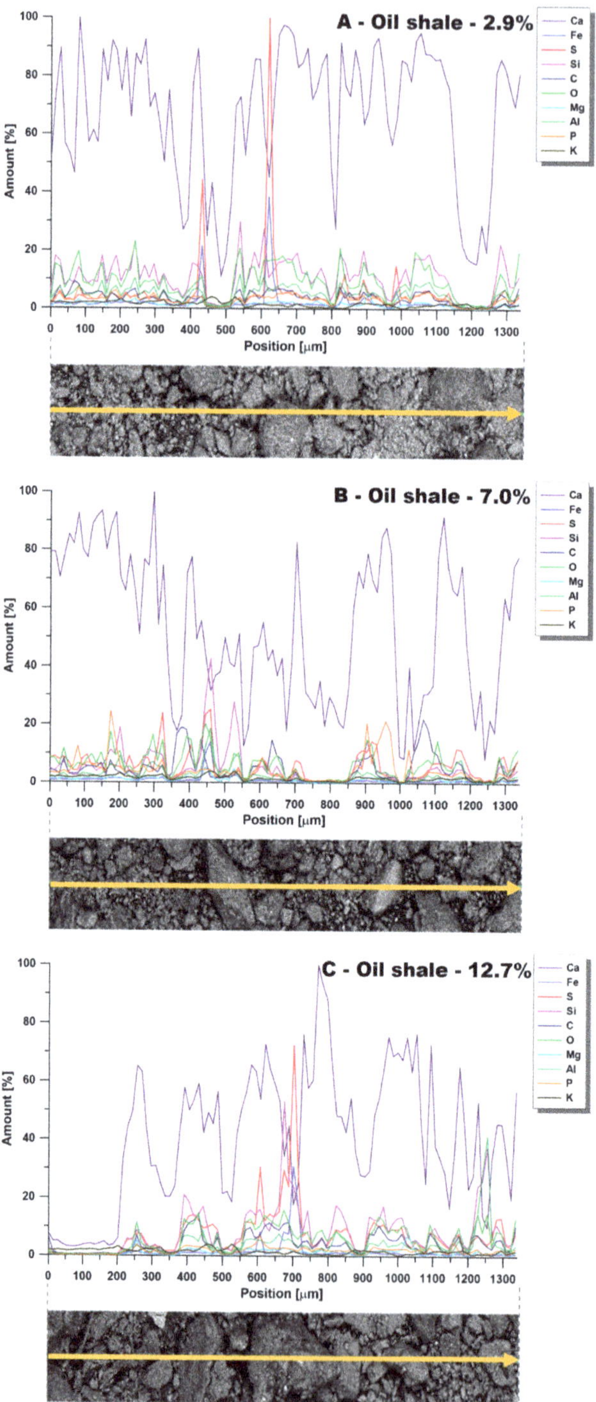

Figure 14. Line analyses of SEM with energy-dispersive X-ray (EDX) detector for **A** (12.5 wt%), **B** (7.0 wt%), and **C** (2.9 wt%) samples.

4. Conclusions

In this work, several high throughput screening techniques for determining shale oil content were applied as an alternative for the conventional Fischer Assay (FA) method. The Omari oil shale deposit, which is located east of Jordan, was characterized and adopted as a case study. Under the test conditions, the following conclusions were drawn:

1. The TGA method could be used as an alternative method to the conventional FA method due to the high linear regression factor (R^2 = 0.99) obtained between the weight loss of the organic content in the pyrolysis range (370–560 °C) obtained by the TGA and the oil yield (wt.%) obtained by the FA method.
2. Elemental analysis could be used for replacing the conventional FA method. A good quadratic correlation (R^2 = 0.98) was maintained between the FA and the hydrogen mass content.
3. The TGA-FTIR method could be used as another quick method for replacing the conventional FA method. A good quadratic correlation (R^2 = 0.97) was maintained between the FA and the aliphatic hydrocarbons (FTIR peak at 2927 cm^{-1}) produced in the pyrolysis zone.
4. XRF analysis of the mineral matter revealed that this oil shale contains mainly calcite and quartz with low concentrations of other metal oxides (Al_2O_3, P_2O_5, Fe_2O_3, MgO, K_2O, and TiO_2).
5. XRD analysis revealed the presence of the following minerals: quartz, ankerite, calcite, apatite, and clay, which matches oil shale composition.
6. SEM analysis revealed an amorphous oil shale surface with no distinct clusters of organic material, which was proven by area mapping and line analysis.

Author Contributions: Conceptualization, Z.A.E.-R., S.A.-G.; data curation, S.A.-G., J.K., Z.A.E.-R.; formal analysis, J.K., S.A.-G., F.Q., Z.A.E.-R.; funding acquisition, Z.A.E.-R.; investigation J.K., S.A.-G., Z.A.E.-R., E.A.; methodology S.A.-G., Z.A.E.-R.; project administration, Z.A.E.-R.; resources, Z.A.E.-R., S.A.-G., J.K.; supervision, Z.A.E.-R., S.A.-G.; validation, Z.A.E.-R., E.A., S.A.-G., J.K., N.A.-R.; visualization, J.K., Z.A.E.-R., E.A., N.A.-R., S.A.-G.; writing—original draft, Z.A.E.-R., S.A.-G., E.A., N.A.-R.; writing—review & editing, Z.A.E.-R., S.A.-G., J.K., N.A.-R.

Funding: This research was funded by the German Jordanian University, project number SAMS 33/2016.

Acknowledgments: The authors would like to thank Jordan Oil Shale Company (JOSCO) for their support in providing the oil shale samples and their Fischer Assay analysis.

Conflicts of Interest: The authors declare no conflict of interest.

References

1. *Energy 2018—Facts and Figures*; Ministry of Energy and Mineral Resources: Amman, Jordan, 2018.
2. Qian, J.; Yin, L. *Oil Shale—Petroleum Alternative*; Petroleum Industry Press: Beijing, China, 2010.
3. Wang, Q.; Hou, Y.; Wu, W.; Liu, Q.; Liu, Z. The structural characteristics of kerogens in oil shale with different density grades. *Fuel* **2018**, *219*, 151–158. [CrossRef]
4. Alali, J.; Salah, A.A.; Yasin, S.M.; Omari, W.A. *Oil Shale*; Ministry of Energy and Mineral Resources: Amman, Jordan, 2015.
5. El-Hasan, T.; Szczerba, W.; Buzanich, G.; Radtke, M.; Riesemeier, H.; Kersten, M. Cr(vi)/cr(iii) and As(v)/As(iii) ratio assessments in Jordanian spent oil shale produced by aerobic combustion and anaerobic pyrolysis. *Environ. Sci. Technol.* **2011**, *45*, 9799–9805. [CrossRef]
6. Alhilu, G. *Jordanian Oil Shale 2019*; Ministry of Energy and Mineral Resources: Amman, Jordan, 2019.
7. Jaber, J.O.; Probert, S.D. Exploitation of Jordanian oil-shales. *Appl. Energy* **1997**, *58*, 161–175. [CrossRef]
8. Zhang, Z.; Yang, X.; Jia, H.; Zhang, H. Kerogen beneficiation from longkou oil shale using gravity separation method. *Energy Fuels* **2016**, *30*, 2841–2845. [CrossRef]
9. Williams, P.F.V. Oil shales and their analysis. *Fuel* **1983**, *62*, 756–771. [CrossRef]
10. Korth, J. Analytical Studies on Australian Oil Shales. Ph.D. Thesis, University of Wollongong, Wollongong, Australia, 1987.

11. Tian, S.; Dong, X.; Wang, T.; Zhang, R.; Zhang, P.; Sheng, M.; Cheng, S.; Zhao, H.; Fei, L.; Street, J.; et al. Surface properties of organic kerogen in continental and marine shale. *Langmuir* **2018**, *34*, 13882–13887. [CrossRef]
12. Foltin, J.P.; Lisboa, A.C.L.; de Klerk, A. Oil shale pyrolysis: Conversion dependence of kinetic parameters. *Energy Fuels* **2017**, *31*, 6766–6776. [CrossRef]
13. Bai, F.; Sun, Y.; Liu, Y.; Guo, M.; Zhao, J. Characteristics and kinetics of Huadian oil shale pyrolysis via non-isothermal thermogravimetric and gray relational analysis. *Combust. Sci. Technol.* **2019**, 1–15. [CrossRef]
14. Hackley, P.C.; Lünsdorf, N.K. Application of raman spectroscopy as thermal maturity probe in shale petroleum systems: Insights from natural and artificial maturation series. *Energy Fuels* **2018**, *32*, 11190–11202. [CrossRef]
15. Lu, Z.Q.; Hai, X.Q.; Wei, J.X.; Bao, R.M. Characterizing of oil shale pyrolysis process with laser ultrasonic detection. *Energy Fuels* **2016**, *30*, 7236–7240. [CrossRef]
16. Budinova, T.; Razvigorova, M.; Tsyntsarski, B.; Petrova, B.; Ekinci, E.; Yardim, M.F. Characterization of Bulgarian oil shale kerogen revealed by oxidative degradation. *Chemie Erde-Geochem.* **2009**, *69*, 235–245. [CrossRef]
17. Stam, A.F.; Meij, R.; te Winkel, H.; van Eijk, R.J.; Huggins, F.E.; Brem, G. Chromium speciation in coal and biomass co-combustion products. *Environ. Sci. Technol.* **2011**, *45*, 2450–2456. [CrossRef] [PubMed]
18. Li, Z.-K.; Wei, X.-Y.; Yan, H.-L.; Zong, Z.-M. Insight into the structural features of zhaotong lignite using multiple techniques. *Fuel* **2015**, *153*, 176–182. [CrossRef]
19. Salmon, E.; Behar, F.; Hatcher, P.G. Molecular characterization of type I kerogen from the green river formation using advanced NMR techniques in combination with electrospray ionization/ultrahigh resolution mass spectrometry. *Org. Geochem.* **2011**, *42*, 301–315. [CrossRef]
20. Kelemen, S.R.; Afeworki, M.; Gorbaty, M.L.; Sansone, M.; Kwiatek, P.J.; Walters, C.C.; Freund, H.; Siskin, M.; Bence, A.E.; Curry, D.J.; et al. Direct characterization of kerogen by x-ray and solid-state ^{13}C nuclear magnetic resonance methods. *Energy Fuels* **2007**, *21*, 1548–1561. [CrossRef]
21. Chu, W.; Cao, X.; Schmidt-Rohr, K.; Birdwell, J.E.; Mao, J. Investigation into the effect of heteroatom content on kerogen structure using advanced ^{13}C solid-state nuclear magnetic resonance spectroscopy. *Energy Fuels* **2019**, *33*, 645–653. [CrossRef]
22. Siramard, S.; Bunman, Y.; Lai, D.; Xu, G. Pyrolysis of Huadian oil shale in an infrared heating reactor. *Energy Fuels* **2017**, *31*, 6996–7003. [CrossRef]
23. Xu, S.; Zeng, X.; Han, Z.; Cheng, J.; Wu, R.; Chen, Z.; Mašek, O.; Fan, X.; Xu, G. Quick pyrolysis of a massive coal sample via rapid infrared heating. *Appl. Energy* **2019**, *242*, 732–740. [CrossRef]
24. Siramard, S.; Zhan, J.-H.; Han, Z.; Xu, S.; Mašek, O.; Xu, G. Secondary cracking of volatile and its avoidance in infrared-heating pyrolysis reactor. *Carbon Resour. Convers.* **2018**, *1*, 202–208. [CrossRef]
25. Luo, X.; Guo, Q.; Zhang, D.; Zhou, H.; Yang, Q. Simulation, exergy analysis and optimization of a shale oil hydrogenation process for clean fuels production. *Appl. Therm. Eng.* **2018**, *140*, 102–111. [CrossRef]
26. Dyni, J.R. *Geology and Resources of Some World Oil-Shale Deposits*; 2005–5294; U.S. Department of the Interior: Washington, DC, USA, 2006; p. 49. Available online: https://pubs.usgs.gov/sir/2005/5294/pdf/sir5294_508.pdf (accessed on 8 August 2019).
27. *Standard Test Method for Oil from Oil Shale (Resource Evaluation by the Fischer Assay Procedure*; ASTM: D 3904-90; ASTM International: West Conshohocken, PA, USA, 1990.
28. Williams, P.F.V. Thermogravimetry and decomposition kinetics of British Kimmeridge clay oil shale. *Fuel* **1985**, *64*, 540–545. [CrossRef]
29. Mingshu, C.; Xiyan, L.; Hongpeng, L.; Yang, X.; Qing, W. Gaseous emission and thermal analysis during co-combustion of oil shale semi-coke and sawdust using TG-FTIR. *Oil Shale* **2015**, *32*, 356–372. [CrossRef]

© 2019 by the authors. Licensee MDPI, Basel, Switzerland. This article is an open access article distributed under the terms and conditions of the Creative Commons Attribution (CC BY) license (http://creativecommons.org/licenses/by/4.0/).

Article

Investigation of Elastomer Seal Energization: Implications for Conventional and Expandable Hanger Assembly

Harshkumar Patel and Saeed Salehi *

Mewbourne School of Petroleum and Geological Engineering, University of Oklahoma, Norman, OK 73071, USA; harsh@ou.edu

* Correspondence: salehi@ou.edu; Tel.: +1-405-325-6829

Received: 16 January 2019; Accepted: 22 February 2019; Published: 25 February 2019

Abstract: Elastomer seals are extensively used in various wellhead and casing/liner hanger equipment as barriers for isolating fluids. Seal assemblies have been identified as one of the major cause of well control incidents. Majority of hangers utilize conventional weight- or mechanical-set slip-and-seal assembly. The objective of this paper is to conduct a detailed investigation of seal energization in conventional and relatively newer expandable type hanger seal assembly. To achieve the objective, the finite element modeling approach was employed. Three dimensional computer models consisting of concentric casings and annular elastomer seal element were constructed. Seal energization process was modelled by manipulating boundary conditions. Conventional seal energization was mimicked by applying rigid support at the bottom of elastomer element and compressing it from the top. Expandable hanger type seal energization was modelled by radially displacing the inner pipe to compress annular seal element. Seal quality was evaluated in terms of contact stress values and profile along the seal-pipe interface. Different amounts of seal energization were simulated. Both types of seal energization processes yielded different contact stress profiles. For the same amount of seal volumetric compression, contact stress profiles were compared. In case of conventional seal energization, contact stress profile decreases from the compression side towards support side. The seal in expandable hanger generates contact stress profile that peaks at the center of contact interface and reduces towards the ends. Convectional seal assembly has more moving parts, making it more prone to failure or under-energization. Finite Element Models were validated using analytical equations, and a good match was obtained. The majority of research related to elastomer seal is focused on material properties evaluation. Limited information is available in public domain on functional design and assessment of seal assembly. This paper adds novel information by providing detailed assessment of advantages and limitations of two different seal energization process. This opens doors for further research in functional failure modes in seal assembly.

Keywords: elastomer seal; finite element analysis; contact pressure; sensitivity analysis; well integrity; liner hanger

---

1. Introduction

Elastomer is a cross-linked network of natural or synthetic polymers. Elastomer material is relatively cheaper and exhibits characteristics property of deforming and recovering under load (elasticity and resilience). Moreover, elastomer seals are suitable for dynamic application [1] and their sealability does not depend on surface characteristic like metal-to-metal seal [2]. Because of these advantages, elastomer material is still widely used in various drilling, completion, and wellhead equipment. In addition to packers, blow-out preventers, and safety valves, elastomer seals is a critical component of casing and liner hanger assemblies.

There are two major types of elastomer seal assembly in liner hangers—conventional and expandable. As shown in Figure 1a, conventional seal assembly is set by applying axial compression load on compression plate or cone either mechanically or hydraulically. As the load is applied, elastomer component expands radially and seals annular pressure below the hanger. Slips element has serrated teeth which engages the opposing surface when the compression is applied and keeps the seal under energization. Expandable liner hanger is a relatively newer technology which offers several operational advantages over traditional assemblies [3–7]. Expandable liner hanger consists of a smooth body with no moving parts and elastomer elements bonded to its outer profile (Figure 1b). The idea is to expand the liner either hydraulically by applying internal pressure or mechanically by running in a solid mandrel having larger outer diameter than the internal diameter of hanger. Expansion of hanger body leads elastomer elements to compress against the casing resulting in seal energization. The seals not only provide hydraulic integrity, but also act as anchor for the liner.

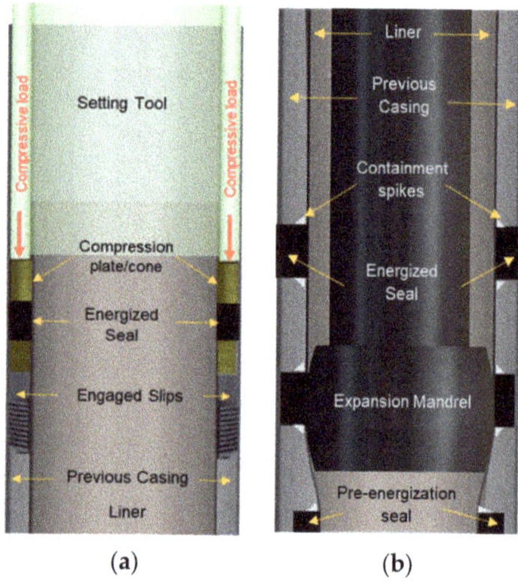

Figure 1. Schematic of (**a**) typical conventional and (**b**) expandable type hanger seal assemblies.

Failure in elastomer seal can compromise the well integrity and lead to a loss of well control (LOWC) event with health, safety, environment, and business consequences. An informal survey of Gulf of Mexico operators conducted by Lohoefer et al. [7] revealed that about 30% to 50% of the seals in liner-casing overlap fail. Another study indicated that up to 18% of offshore wells have some form of failure or weakness [8]. This problem has also been acknowledged by regulator Bureau of Safety and Environmental Enforcement (BSEE) in a recent investigation report of an incident [9]. Analysis of 156 LOWC events' data (2000–2015) [10,11] reveal that almost half (46%) of the causes of secondary barrier failure are associated with seal or seal containing components of the well (Figure 2). A 2015 survey by Oil & Gas iQ [12] puts seals as one of the top technical challenges (18% of total) associated with High-Pressure High-Temperature (HPHT) oil and gas exploration.

Majority of the research related to elastomer seal and industry standards/guidelines have been focused on material composition, properties evaluation, and qualification [11,13]. Elastomer seal failure in the form of physical and chemical degradation such as rapid gas decompression, extrusion, compression set, abrasion, volumetric swelling, etc., is widely studied [11,14–16]. However, very limited information is available on function aspects of elastomer seal assemblies. Particularly, major research gaps are: the effect of seal energization method, the impact of energization failure, the consequence of failure in supporting components, and the effect of low quality energization. This paper attempts to fill one of these knowledge gaps by performing detailed comparison of conventional versus expandable type energization method.

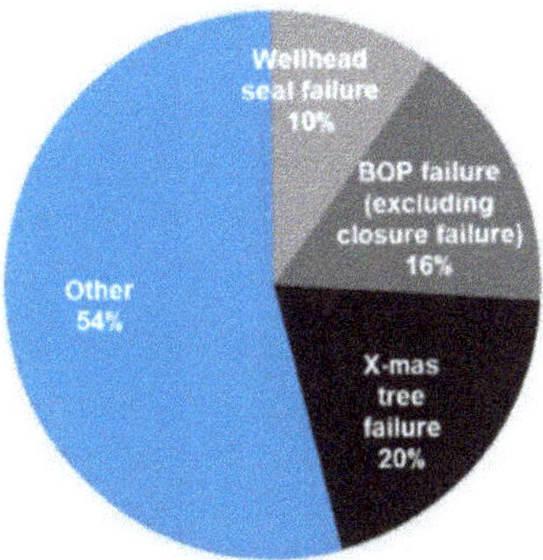

Figure 2. Causes of secondary barrier failure in 156 loss of well control events occurred during 2000–2015.

2. Literature Review

Commonly used elastomer material in the oil & gas industry can be grouped into seven groups—NBR (Nitrile Butadiene Rubber), HNBR (Hydrogenated Nitrile Butadiene Rubber), EPDM (Ethylene Propylene Diene Monomer), FKM (Fluorocarbon), FEPM (Tetrafluoroethylene Propylene), FFKM (Perfluorocarbon), and PTFE (Polytetrafluoroethylen). The elastomer used are relatively stiffer and strain values are less, permitting assumption of linear elastic material behavior requiring only elastic modulus and Poisson's ratio [17]. For high deformations, use of hyperelastic material model such as neo-Hookean, Mooney–Rivlin, Ogden, Yeoh, and so forth, is needed. The hyperelastic material model requires various physical measurements like uniaxial, planer, biaxial stress behavior, and volumetric compression data.

The majority of relevant studies in the literature are focused on packer equipment, and to some extent, expandable tubular. Nonetheless, they provide useful information. Berger [18] performed physical tests on 7 $\frac{3}{4}$-in. packer to assess different types of backup mechanisms such as steel foldback ring, mesh rings, and garter springs that support seal during energization process and maintains it under compression. Feng et al. [19] conducted 2D Finite Element Analysis (FEA) of axial compression in packer equipment with two elastomer components. They observed consistently higher contact pressure at the top elastomer element compared to bottom elastomer element. Alzebdeh et al. [20] studied the effect of seal length, thickness, and compression amount in 2D FEA model of expandable tubular sealing against different formations. Al-Kharusi et al. [21] and Al-Hiddabi et al. [22] conducted theoretical analysis of compression of elastomer seal in expandable tubular. Assuming linear elastic material properties, they presented analytical model to predict contact stress as a function of different amount of compression and differential fluid pressure across the seal.

Lin [23] used FEA to investigate structural integrity of slip element of packer. The author modelled different teeth spacing in slip and examined how it affects risk of failure. Ma et al. [24] used 2D FEA model of swelling elastomer packer element to investigate the effects of seal length, swelling amount, and different formation on seal quality and contact stress. They observed that in swellable elastomer packer equipment, contact stress profile peaks at the center of the element and declines towards both ends of the axial length. Wang et al. [25] studied elastomer failure modes such as extrusion, sliding, and rupture by conducting theoretical analysis supported by visual experimental observations. Zhong et al. [26] used FEA model of large bore expandable liner hanger to assess expansion force, cone pull out fore, contact stress at the seal-pipe interface, and deformations in hanger, casing, containment spikes, and cone body.

Hu et al. [27] conducted 3D finite element analysis of compression packer equipment. They studied the effect of carbon black content in HNBR formulation and ranked the materials in terms of contact stress, shoulder extrusion, and stresses in structure of the packer. Elhard et al. [11] conducted extensive material properties measurement of commonly used oil and gas elastomer material and presented hyper-elastic models parameters. They also conducted experimental and finite element modelling study of O-ring extrusion. Recently, Patel et al. [13] used 3D finite element modelling to evaluate the performance of conventional liner hanger seal assembly. They conducted parametric study and ranked various design parameters based on the impact on seal performance. They also presented an empirical correlation to predict contact pressure.

3. Objectives and Scope

The objective of this paper was to compare seal energization and resultant contact stress profiles in conventional and expandable type hanger seal assembly.

For this purpose, three dimensional finite element model consisting of liner, elastomer seal, and casing elements was created. Both types of energization were simulated by manipulating boundary conditions. The simulations were performed at different amount of volumetric compressions. Comparison between both types of equipment was performed in terms of contact stress profile generated at the seal-pipe interface. For thorough comparison, investigation was conducted using both linear-elastic and hyper-elastic material models under frictionless and frictional surface conditions. For expandable hanger, configuration of containment spikes were also varied to examine its impact on contact stress profile. The commonly used FKM elastomer was employed as the reference material in the study.

4. Finite Element Models

Two dimensional schematic and top view of the both FEA models are shown in Figure 3. Dimensions of 18-in. liner and 20-in. casing were based on the actual well design where cement in liner-casing overlap likely failed [9]. The length of pipes were kept long enough to avoid any end-effects. Seal axial length was 2.5-in. and radial width was 0.6875-in. To save number of mesh nodes used and minimize computational power, 1/16th of the model was used for the simulations.

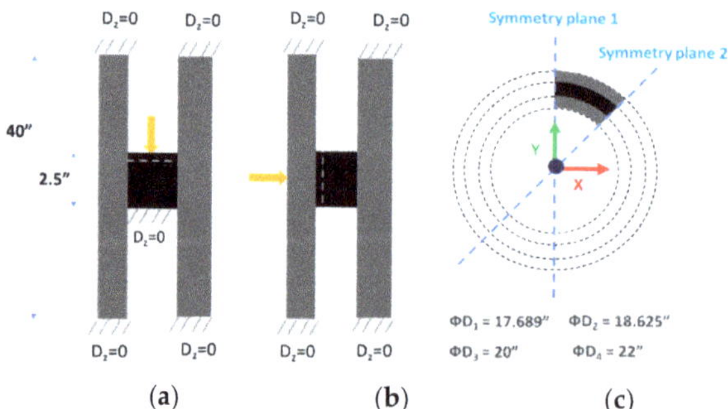

Figure 3. Two dimensional schematic of the Finite Element Analysis (FEA) models of (**a**) conventional and (**b**) expandable type hanger seal assembly, and (**c**) the top view.

Conventional seal energization was mimicked by applying zero displacement boundary conditions i.e., support at the bottom of elastomer element and compressing it from the top. To achieve compression, displacement type boundary condition was used. Displacement boundary condition was used instead of directly applying load because it provides faster and more controlled numerical convergence with less susceptibility to failure. Expandable hanger type seal energization was modelled by radially displacing the liner to compress annular seal element. Casing was fixed radial and axial during the energization process.

Liner and casing were modelled as isotropic linear elastic material. FKM elastomer was modelled as both linear elastic and hyper-elastic (Ogden 3rd order) material. Material properties of FKM were obtained from existing literature [11]. FKM is a common elastomer material used in the oil and gas applications because of its high resistance to chemical degradation [28]. The material properties for all three components are listed in Table 1. Hyper-elastic stress-strain behavior of FKM is graphically presented in Figure 4.

Table 1. Material properties used in the FEA model.

Material	Properties
Linear elastic FKM at 73°F [11]	Young's modulus = 310.5 psi Poisson's ratio = 0.49
Hyper-elastic FKM at 73°F [11]	Ogden 3rd Order μ_1 = 278 psi, μ_1 = 32.31 psi, μ_1 = 0.198 psi α_1 = 2.661, α_1 = -2.661, α_1 = 10.79 D1 = 1.4×10^{-5} psi^{-1}, D_2 = 2.7×10^{-6} psi^{-1}, D_1 = 0
Liner and casing	Young's modulus = 29×10^6 psi Poisson's ratio = 0.3

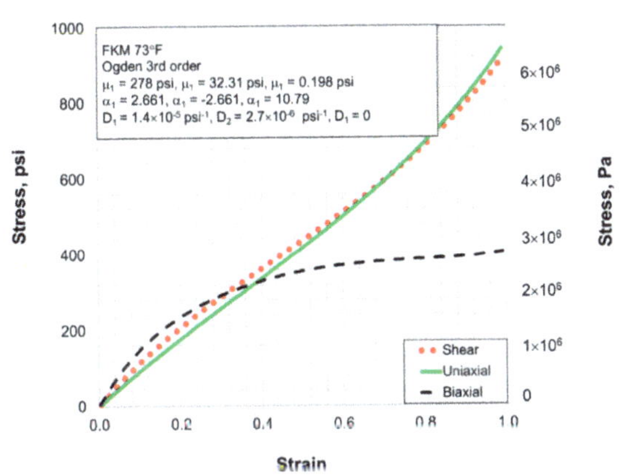

Figure 4. Uniaxial, shear, and biaxial material behavior of Fluorocarbon (FKM).

The model was discretized using hexahedral mesh because of its near orthogonality. Augmented Lagrange type contact formulation was used for representing seal-pipe contact interfaces. Stiffness factor was adjusted to achieve minimum possible penetration (10^{-3} to 10^{-4} inch) while still successfully achieving convergence.

Model verification was performed by ensuring that boundary conditions were being fulfilled and results were independent of symmetry plane selection and mesh size. Additionally, analytical validation was performed to ensure accuracy of the model predictions.

5. Analytical Validation

To validate the FEA model, the analytical relationship between bulk modulus, volumetric compression, and pressure can be used. As shown in the Figure 5, conventional hanger seal model is constrained in radial and axial direction after energization. The pressure generated at all four frictionless contacting surfaces should be same. This situation is similar to how bulk modulus is defined, i.e. application of equal external pressure over the surface of a three dimensional body to achieve bulk volumetric compression.

$$\frac{\Delta V}{V} = -\frac{P}{K}. \tag{1}$$

$$K = \frac{E}{3(1-2\nu)} \tag{2}$$

where P, K, and ν are pressure, bulk modulus, and Poisson's ratio, respectively. V is the original volume of elastomer seal, and ΔV is change in volume as shown in Figure 5.

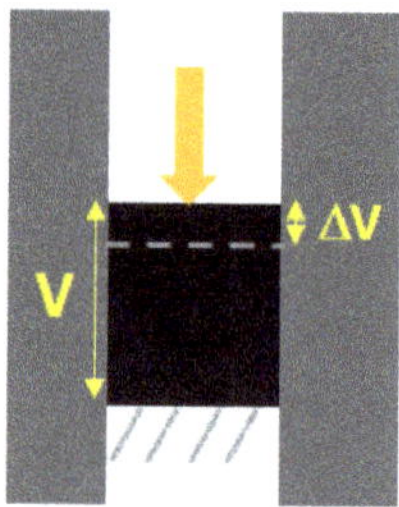

Figure 5. Volumetric compression in FEA model of conventional hanger seal.

Pressure calculated using Equation (1) were compared with FEA simulated contact pressure for different volumetric compression values. As shown in the Figure 6, a good match was obtained. Deviation from analytical calculation was 4% to 7% at lower ends of compression (less than 5%) and less than 3% at higher ends of compression (more than 5%).

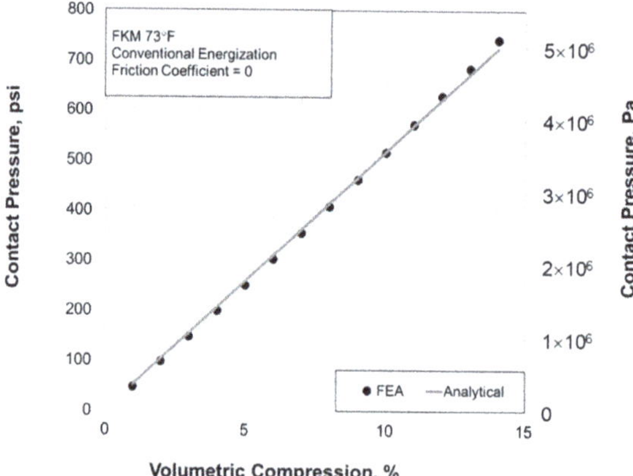

Figure 6. Comparison between contact stress simulated by FEA model and calculated using analytical equation.

6. Simulation Results

Five parameters were varied to run a total of 50 simulation cases. Factors investigated were energization method, amount of volumetric compression, friction coefficient, and material behavior. Results were grouped based on the individual parameters being examined.

6.1. Effect of Energization Method

Contact stress profiles generated in conventional and expandable liner hanger are presented in Figures 7 and 8 respectively. The profiles were generated at different volumetric compression using linear elastic FKM elastomer. Compression ratio defined as % change in axial seal height (for conventional assembly) and % change in seal inner radius (for expandable assembly) are also provided. Surface conditions were assumed to be frictionless. For expandable seal assembly, elastomer containment in axial direction was not considered.

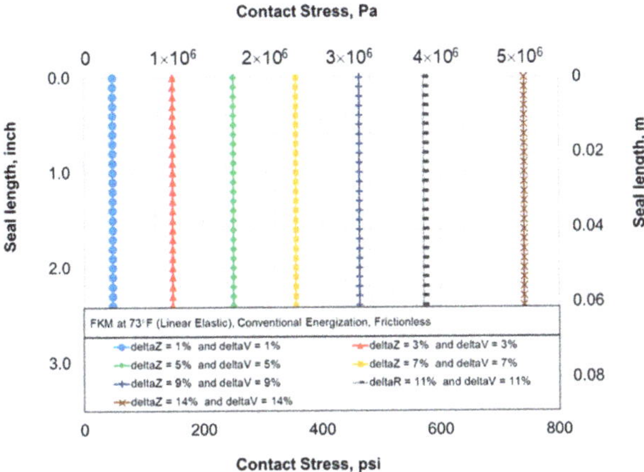

Figure 7. Contact stress profiles generated in conventional seal assembly at different volumetric compression.

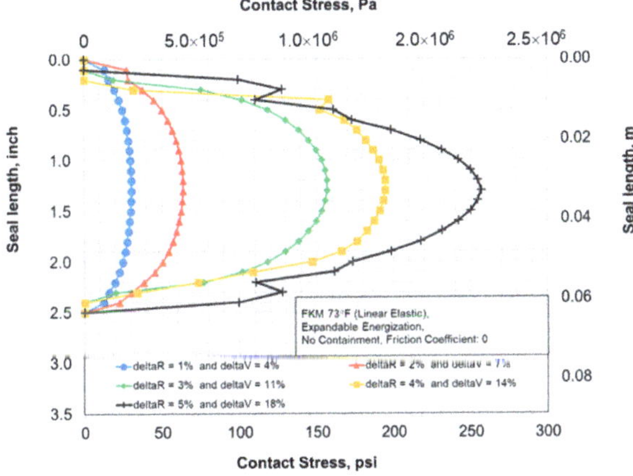

Figure 8. Contact stress profiles generate in expandable seal assembly at different volumetric compression.

As shown in the Figure 7, contact stress values remained constant along the seal length. This is because of the frictionless assumption for surface. For expandable liner hanger seal assembly (Figure 8), the contact stress peaked at the center of the seal and declines away from it. This is due to the fact that elastomer is not contained at the ends. Plus, seal-pipe interface is frictionless. Hence, the seal slides away from the center while being compressed; leading to peak contact stress at the middle and zero contact stress at the ends.

It is clear from the profiles that contact stress value increases with an increase in the amount of compression. The increment is practically linear for both of the seal assemblies as shown in the Figure 9. For the same amount of volumetric compression, conventional energization yields higher contact stress. For example, at 7% volumetric compression, conventional and expandable assembly generate 355 psi and 63 psi peak contact stress, respectively. This is expected because of the lack of elastomer containment in expandable seal assembly.

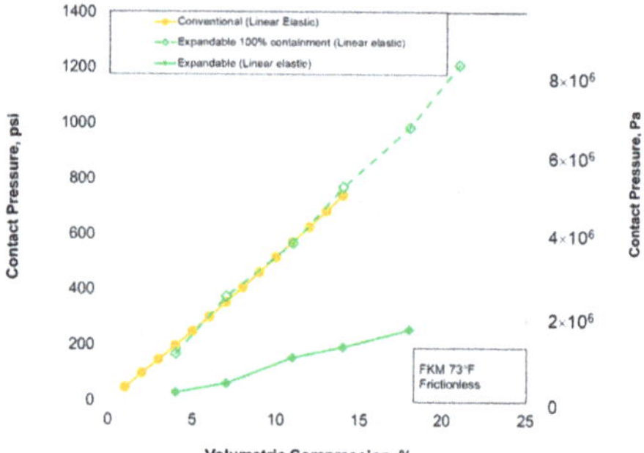

Figure 9. Contact pressure as a function of volumetric compression for conventional and expandable assembly (with and without elastomer containment).

Next, the effect of elastomer containment spikes was studied. Containment spikes were represented by zero displacement boundary conditions as shown in Figure 10. Five containment configurations were studied—0, 25, 50, 75, and 100% containment. The configurations on either side were kept the same. Volumetric compression was kept constant at 4% and surfaces were considered frictionless.

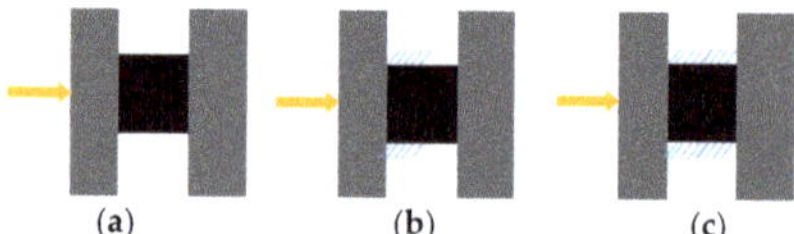

Figure 10. Elastomer seal containment in expandable liner hanger: (**a**) no containment, (**b**) 50% containment, and (**c**) 100% containment.

It is clear from the results shown in Figure 11 that as elastomer containment increases, contact stress profile changes from parabolic to a progressively flatter one and eventually becomes constant at 100% containment. Peak contact stress values increases with increase in % containment. The increment is not linear. At the same volumetric compression contact stress profile and values at 100% containment is same as in the conventional seal assembly (Figures 9 and 12).

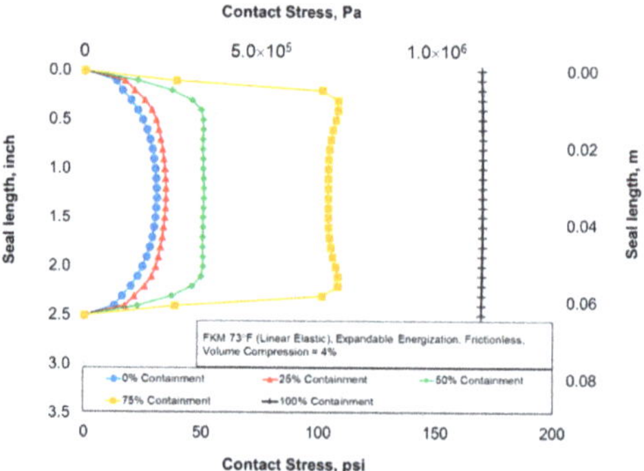

Figure 11. Contact stress profiles generated in expandable seal assembly at different amounts of seal containment.

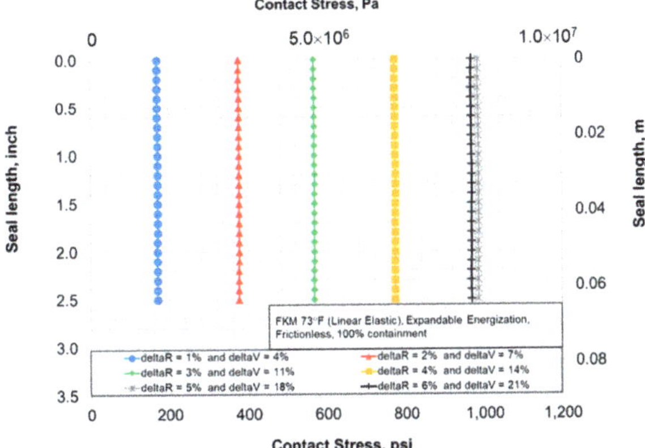

Figure 12. Contact stress profiles generate in expandable seal assembly with 100% containment and different volumetric compression.

6.2. Effect of Material Behavior

Next, the effect of material behavior was studied. Specifically, hyper-elastic material behavior of FKM (Ogden 3rd order) was modelled and resultant contact stress profiles were compared with the ones generated using linear-elastic material behavior. Contact stress profiles using the hyper-elastic FKM for conventional and expandable assembly without containment are presented in Figures 13 and 14 respectively.

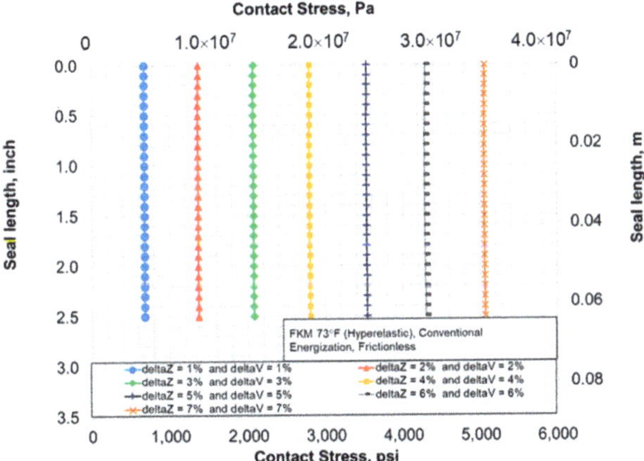

Figure 13. Contact stress profiles in conventional seal assembly for hyper-elastic FKM.

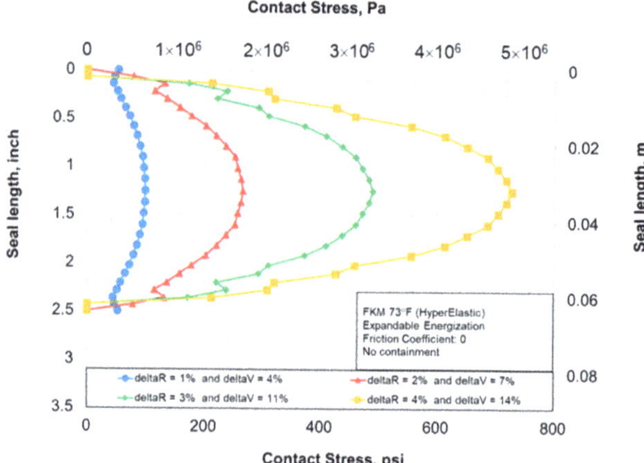

Figure 14. Contact stress profiles in expandable seal assembly for hyper-elastic FKM.

Results indicate that the contact stress profile in both assembly remains unchanged after switching to hyper-elastic material model. The contact stress values are notably higher than in the case of linear-elastic model (Figure 15). This is most likely due to the fact that in hyper-elastic model, elastomer material typically becomes stiffer at higher strain values and hence, at the same amount of deformation, higher contact stress is generated. As shown in the Figure 15, seal energization curve still remained practically linear with hyper-elastic material behavior.

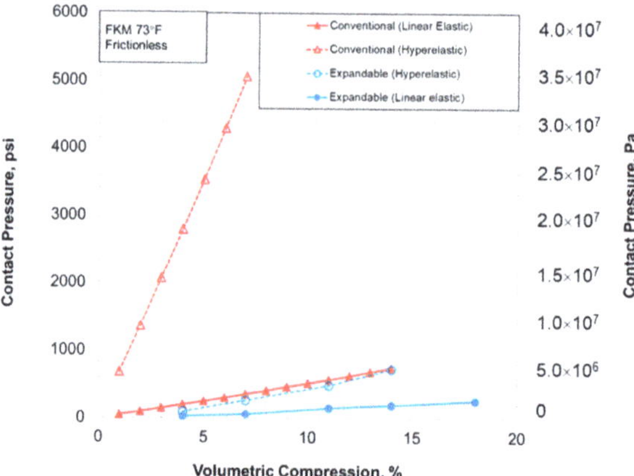

Figure 15. Contact pressure as a function of volumetric compression for conventional and expandable assembly with linear- and hyper-elastic material behavior.

6.3. Effect of Friction

All results discussed so far were generated assuming frictionless contact surface. Our next task was to investigate effect of friction on contact stress profiles generated in conventional and expandable seal assemblies. Typical friction coefficient between elastomer and steel tubing is 0.3 or higher [29].

Figure 16 presents contact stress profiles in conventional seal assembly assuming friction coefficient of 0.3 at elastomer-pipe interface. Frictionless contact stress profiles are also overlapped for easier comparison. In the presence of friction, contact stress profile is no longer constant. At the compression side of the seal, contact stress values are higher than the frictionless reference. Conversely, at the support side, contact pressure values are notably lesser than frictionless. It is clear that the compression or energization is not translated to the axially opposite end of the seal. The effect is more pronounced as volumetric compression increases. It can also be extrapolated that longer seal would also have more pronounced effect of friction. Failure to consider true frictional effect in conventional seal assembly design could be detrimental since significantly lower contact pressure values at the bottom of the seal can increase the risk of fluid penetration.

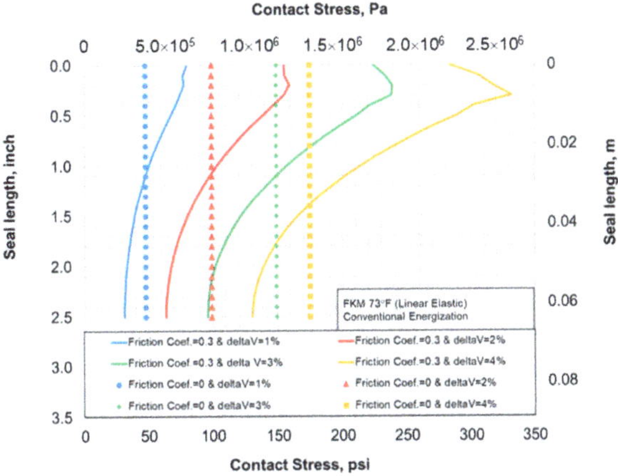

Figure 16. Effect of friction on contact stress profile in conventional seal assembly.

As shown in Figure 17, unlike conventional seal assembly, expandable assembly retained the overall shape of contact stress profile in presence of frictional sliding. However, the peak contact stress value increased with increase in friction coefficient from 0 to 0.3 to 0.6.

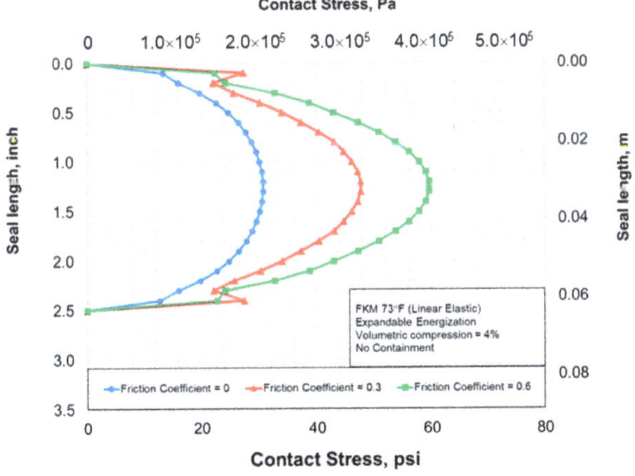

Figure 17. Effect of friction on contact stress profile in expandable seal assembly.

7. Practical Implications

For the same volumetric compression, conventional seal assembly yields higher contact pressure than expandable seal assembly without containment. At 100% elastomer containment, contact stress profile in the expandable assembly matches with the conventional assembly. Actual expandable seal hanger assembly would almost always have elastomer containment spikes on either sides. However, to facilitate smooth running-in of the tool into the well, the spikes have to be shorter. This would provide less than 100% containment. Thus, in frictionless conditions, assuming that supporting components of either assemblies do not fail, conventional type energization will always yield higher contact pressure and should be preferred to the expandable type energization mechanism.

However, in the real field applications, contact surfaces are never frictionless. In presence of frictional stresses, contact stress profile of conventional seal assembly deviates from the uniform profile observed in frictionless condition (Figure 16). Moreover, the deviation is significantly dependent on amount of volumetric compression. Contact pressure peaks at a shorter distance from the compression side and declines rapidly towards the support side (Figure 16). Low contact pressure at the bottom would permit easy penetration of the fluid. On the other side, in expandable type energization, contact stress profile in frictional case maintains the same symmetry as the frictionless case (Figure 17). The profile becomes slightly narrower at the center but the contact pressure values at all locations are higher than the frictionless case. This improves the seal's performance. Although, peak contact pressure values are lesser than in the conventional energization, the difference would reduce with elastomer containment spikes. Overall, in presence of friction, expandable type energization is likely to be more reliable than the conventional energization because the former yields uniformly higher contact pressure than frictionless case while also maintaining the symmetry of the contact pressure profile.

Conventional seal assembly, in addition to the compression plates on either side, has moving parts like slips for supporting bottom plate and a mechanism to exert load on the top plate. Expandable seal assembly only has non-moving containment spikes on either side of the seal (Figure 1). If elastomer containment spikes in expandable assembly fail, the seal would still maintain contact pressure as shown in Figure 8. However, if slips or compression plates in conventional assembly fail, contact pressure and consequently sealability will be completely lost.

Expandable type seal energization also has additional design advantages over the conventional assembly. In expandable assembly, the seal is energized radially. Hence, it is possible to install multiple seal elements along the length of pipe and achieve same contact pressure in all of them. This redundancy further minimizes the risk of failure. In conventional assembly with multiple alternating seal elements and compression plates, because of the frictional stress being parallel to the compression load, it is not possible to achieve the same contact pressure in each seal. Based on the effect of friction shown in Figure 16, it can be inferred that the peak contact pressure would subsequently decrease from the top seal element to the bottom seal element. This effect of friction has also been demonstrated previously by Ma et al. [27] in simulations of dual rubber packer equipment.

As discussed in Section 6.2, selection of material model in modelling did not impact the shape of contact pressure profile in either of the seal assemblies. However, hyper-elastic FKM yielded higher contact pressure values than the linear elastic FKM. This is because of the fact that FKM material exhibited more stiffness at higher strain values. Thus, to prevent under-estimation or over-estimation of the seal's performance, it is important to measure the elastomer material behavior over the range of operating strains and use correct material model in predictions.

8. Conclusions

Using three dimensional finite element models, this study presents detailed comparison of seal energization in conventional and expandable liner hanger seal assemblies. The following are major conclusions from this work:

- In conventional seal assembly, contact stress value decreases along the seal length from the compression side towards the support side. In case of frictionless assumption, contact stress remains constant along the seal length.
- In expandable seal assembly, irrespective of friction coefficient, contact stress peaks at the center of the seal length and declines towards either sides of the axial ends. The profile becomes progressively flatter with increase in elastomer seal containment and becomes similar to conventional seal assembly at 100% containment.

- Contact pressure values increases with increase in amount of compression, i.e., %volumetric compression. The increment is practically linear, irrespective of energization method and material behavior.
- Selection of material model in modelling did not impact the shape of contact pressure profile in either of the seal assemblies. However, hyper-elastic FKM yielded higher contact pressure values than the linear elastic FKM. Therefore, it is important to measure the elastomer material behavior over the range of operating strains and use appropriate model in predictions.
- In frictionless condition, conventional type energization will almost always provide higher peak contact pressure values and should be preferred to the expandable type energization mechanism.
- In case of frictional contacts, expandable type energization is likely to be more reliable than the conventional energization because the former yields higher contact pressure than the frictionless case while also maintaining the symmetry of the contact pressure profile.
- Expandable energization is more robust to failure in supporting components than the conventional assembly. Even if both elastomer containment spikes completely fail, the expandable seal assembly would still maintain contact pressure.
- In expandable energization, it is possible to install multiple seal elements along the length of pipe and achieve same contact pressure in all elements. In conventional assembly with multiple alternating seals and compression plates, peak contact pressure would subsequently decrease from the top seal element to the bottom seal element.

Author Contributions: H.K.P. conceptualized the idea, developed FEA models, and ran simulations. S.S. provided guidance in data analysis and paper writeup.

Funding: Work was partially funded by Bureau of Safety and Environmental Enforcement (Project No: E17PC00005).

Acknowledgments: The authors would like to extend their sincere gratitude to the University of Oklahoma for providing necessary resources and granting the permission to publish this work.

Conflicts of Interest: The authors declare no conflict of interest.

Nomenclature

Acronyms	
BSEE	Bureau of Safety and Environmental Enforcement
EPDM	Ethylene Propylene Diene Monomer
FEPM	Tetrafluoroethylene Propylene
FFKM	Perfluorocarbon
FEA	Finite Element Analysis
FKM	Fluorocarbon
HPHT	High Pressure High Temperature
HNBR	Hydrogenated Nitrile
LOWC	Loss of Well Control
Symbols	
α_i	Ogden 3rd order material constant
deltaR	Percentage change in inner radius of seal
deltaV	Change in seal volume due to energization
deltaZ	Percentage change in axial seal height
D_i	Ogden 3rd order material constant
E	Elastic modulus / Young's modulus
K	Bulk modulus
μ_i	Ogden 3rd order material constant
ν	Poisson's ratio
P	Contact pressure
V	Initial seal volume
ΔV	Change in seal volume due to energization

References

1. Tu, B.; Cheng, H.L. Alternative Methodology for Elastomeric Seal RGD and Aging Testing Validates Long-Term Subsea Seal Performance and Integrity. In Proceedings of the Offshore Technology Conference, Houston, TX, USA, 2–5 May 2016. [CrossRef]
2. Patel, H.; Hariharan, H.; Bailey, G.; Jung, G. Advanced computer modelling for metal-to-metal seal in API flanges. In Proceedings of the SPE Annual Technical Conference and Exhibition, Dallas, TX, USA, 24–26 September 2018. [CrossRef]
3. Mullins, F. Metal-Formed Liner Hanger Avoids High-Setting-Pressure Requirements. *J. Pet. Technol.* **2016**, *68*, 4–8. [CrossRef]
4. Mccormick, J.; Matice, M.; Cramp, S. Big Bore Expandable Liner Hangers for Offshore and Deepwater Applications Reduces Cost and Increases Reliability: Global Case History. In Proceedings of the SPETT 2012 Energy Conference and Exhibition, Port-of-Spain, Trinidad, 11–13 June 2012. [CrossRef]
5. Walvekar, S.; Jackson, T. Development of an Expandable Liner-Hanger System to Improve Reliability of Liner Installations. In Proceedings of the Offshore Technology Conference, Houston, TX, USA, 30 April–3 May 2007. [CrossRef]
6. Smith, P.; Williford, J. Case Histories: Liner-Completion Difficulties Resolved With Expandable Liner-Top Technology. In Proceedings of the Canadian International Petroleum Conference, Calgary, AB, Canada, 13–15 June 2006. [CrossRef]
7. Lohoefer, C.L.; Mathis, B.; Brisco, D.; Waddell, K.; Ring, L.; York, P. Expandable liner hanger provides costeffective alternative solution. In Proceedings of the IADC/SPE Drilling Conference, New Orleans, LA, USA, 23–25 February 2000. [CrossRef]
8. Van Dort, R. Metal-to-metal seals meet downhole hazard demands. *J. Pet. Technol.* **2009**, *61*, 24–26. [CrossRef]
9. BSEE. *QC-FIT Evaluation of Seal Assembly & Cement Failures Interim Summary of Findings*; Internal QC-FIT Report #2014-02; Bureau of Safety and Environmental Enforcement: Washington, DC, USA, 2014.
10. Holand, P. *Loss of Well Control Occurrence and 49 Size Estimators, Phase I and II*; 51 Report #ES201471/2; Bureau of Safety and Environmental Enforcement: Washington, DC, USA, 2017.
11. Elhard, J.D.; Duguid, A.; Heinrichs, M. Research on Safety Technology Verification for Materials and Pressure High Temperature (HPHT) Continental Shelf (OCS), High Corrosions in the U.S. Outer Material Evaluation. Technical Assessment Program Report (TAP 767AA) Prepared for Bureau of Safety and Environmental Enforcement. Available online: https://www.bsee.gov/tap-technical-assessment-program/research-on-safety-technology-verification-for-materials-and (accessed on 15 September 2018).
12. Oil & Gas iQ. High Pressure High Temperate, High Costs, High Stakes? 2015. Available online: https://www.oilandgasiq.com/content-auto-download/5b04c1b543dfd0385d3c7c22 (accessed on 4 October 2018).
13. Patel, H.; Salehi, S.; Teodoriu, C.; Ahmed, R. Performance evaluation and parametric study of elastomer seal in conventional hanger assembly. *J. Pet. Sci. Eng.* **2019**, *175*, 246–254. [CrossRef]
14. Fernández, C.; Castaño, P. Compatibility Behavior of Elastomers for PCP Applications. 2016. Available online: https://www.onepetro.org/conferencepaper/NACE-2016-7106 (accessed on 20 October 2018).
15. Cong, C.B.; Cui, C.C.; Meng, X.Y.; Lu, S.J.; Zhou, Q. Degradation of hydrogenated nitrile-butadiene rubber in aqueous solutions of H_2S or HCl. *Chem. Res. Chin. Univ.* **2013**, *29*, 806–810. [CrossRef]
16. Campion, R.P.; Thomson, B.; Harris, J.A. Elastomers for Fluid Containment in Offshore Oil and Gas Production: Guidelines and Review. Research Report (RR 320) Prepared by MERL Ltd for the Health and Safety Executive, U.K. 2005. Available online: http://www.hse.gov.uk/research/rrpdf/rr320.pdf (accessed on 20 September 2018).
17. Bosma, M.G.R.; Cornelissen, E.K.; Schwing, A. Improved Experimental Characterization of Cement/Rubber Zonal Isolation Materials. In Proceedings of the SPE Asia Pacific Oil and Gas Conference and Exhibition, Brisbane, Australia, 16–18 October 2000. [CrossRef]
18. Berger, S. Experimental and Finite Element Analysis of High Pressure Packer Elements. Master's Thesis, MA Institute of Technology, Cambridge, MA, USA, September 2004.
19. Feng, D.; Yuan, Y.; Tan, B.; Yang, C.; Xu, G.; Wang, P. Finite Element Analysis of The Packer Rubbers on Sealing Process. In Proceedings of the 2010 International Conference on Mechanic Automation and Control Engineering, Wuhan, China, 26–28 June 2010. [CrossRef]

20. Alzebdeh, K.; Pervez, T.; Qamar, S.Z. Finite Element Simulation of Compression of Elastomeric Seals in Open Hole Liners. *J. Energy Resour. Technol.* **2010**, *132*, 031002. [CrossRef]
21. Al-Kharusi, M.S.; Qamar, S.Z.; Pervez, T.; Akhtar, M. Non-Linear Model for Evaluation of Elastomer Seals Subjected to Differential Pressure. In Proceedings of the SPE/DGS Saudi Arabia Section Technical Symposium and Exhibition, Al-Khobar, Saudi Arabia, 15–18 May 2011. [CrossRef]
22. Al-Hiddabi, S.A.; Pervez, T.; Qamar, S.Z.; Al-Jahwari, F.K.; Marketz, F.; Al-Houqani, S.; van de Velden, M. Analytical model of elastomer seal performance in oil wells. *Appl. Math. Model.* **2015**, *39*, 2836–2848. [CrossRef]
23. Lin, Z.C. The strength analysis and structure optimization of packer slip based on ANSYS. *Appl. Mech. Mater.* **2013**, *423*, 1967–1971. [CrossRef]
24. Ma, M.; Jia, W.; Bu, Y.; Guo, S. Study on rubber seal design of a swellpacker in oil well cementing. *Open Access Libr. J.* **2014**, *1*, 1. [CrossRef]
25. Wang, Z.; Chen, C.; Liu, Q.; Lou, Y.; Suo, Z. Extrusion, slide, and rupture of an elastomeric seal. *J. Mech. Phys. Solids* **2017**, *99*, 289–303. [CrossRef]
26. Zhong, A.; Johnson, M.R.; Kohn, G.; Koons, B.; Saleh, M. Performance Evaluation of a Large Bore Expandable Liner Hanger for Field Operations in the Gulf of Mexico. In Proceedings of the Offshore Technology Conference, Houston, TX, USA, 4–7 May 2015. [CrossRef]
27. Hu, G.; Zhang, P.; Wang, G.; Zhang, M.; Li, M. The influence of rubber material on sealing performance of packing element in compression packer. *J. Nat. Gas Sci. Eng.* **2017**, *38*, 120–138. [CrossRef]
28. Salehi, S.; Ezeakacha, C.P.; Kwatia, G.; Ahmed, R.; Teodoriu, C. Performance verification of elastomer materials in corrosive gas and liquid conditions. *Polym. Test.* **2019**, *75*, 48–63. [CrossRef]
29. Ma, W.; Qu, B.; Guan, F. Effect of the friction coefficient for contact pressure of packer rubber. *Proc. Inst. Mech. Eng. Part C J. Mech. Eng. Sci.* **2014**, *228*, 2881–2887. [CrossRef]

© 2019 by the authors. Licensee MDPI, Basel, Switzerland. This article is an open access article distributed under the terms and conditions of the Creative Commons Attribution (CC BY) license (http://creativecommons.org/licenses/by/4.0/).

MDPI
St. Alban-Anlage 66
4052 Basel
Switzerland
Tel. +41 61 683 77 34
Fax +41 61 302 89 18
www.mdpi.com

Energies Editorial Office
E-mail: energies@mdpi.com
www.mdpi.com/journal/energies

www.ingramcontent.com/pod-product-compliance
Lightning Source LLC
LaVergne TN
LVHW070949170726
843515LV00005B/950

* 9 7 8 3 0 3 9 2 8 8 7 5 5 *